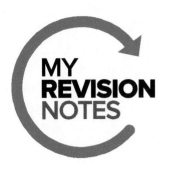

MY REVISION NOTES

City & Guilds

Level 2 Advanced Technical Diploma (8202-20)

ELECTRICAL INSTALLATION

Peter Tanner

HODDER
EDUCATION
AN HACHETTE UK COMPANY

Acknowledgements

We would like to thank City & Guilds for permission to reuse artworks from their Electrical Installations textbooks.

Every effort has been made to trace all copyright holders, but if any have been inadvertently overlooked, the Publishers will be pleased to make the necessary arrangements at the first opportunity.

Although every effort has been made to ensure that website addresses are correct at time of going to press, Hodder Education cannot be held responsible for the content of any website mentioned in this book. It is sometimes possible to find a relocated web page by typing in the address of the home page for a website in the URL window of your browser.

Hachette UK's policy is to use papers that are natural, renewable and recyclable products and made from wood grown in well-managed forests and other controlled sources. The logging and manufacturing processes are expected to conform to the environmental regulations of the country of origin.

Orders: please contact Hachette UK Distribution, Hely Hutchinson Centre, Milton Road, Didcot, Oxfordshire, OX11 7HH. Telephone: +44 (0)1235 827827. Email education@hachette.co.uk Lines are open from 9 a.m. to 5 p.m., Monday to Friday. You can also order through our website: www.hoddereducation.co.uk

ISBN: 978 1 3983 2734 4

Cover photo © krasyuk – stock.adobe.com

Typeset in India.

Printed and bound by CPI Group (UK) Ltd, Croydon, CR0 4YY

A catalogue record for this title is available from the British Library.

Get the most from this book

Everyone has to decide his or her own revision strategy, but it is essential to review your work, learn it and test your understanding. These Revision Notes will help you to do that in a planned way, topic by topic. Use this book as the cornerstone of your revision and don't hesitate to write in it – personalise your notes and check your progress by ticking off each section as you revise.

Tick to track your progress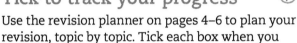

Use the revision planner on pages 4–6 to plan your revision, topic by topic. Tick each box when you have:
+ revised and understood a topic
+ tested yourself
+ practised the exam questions and gone online to check your answers.

You can also keep track of your revision by ticking off each topic heading in the book. You may find it helpful to add your own notes as you work through each topic.

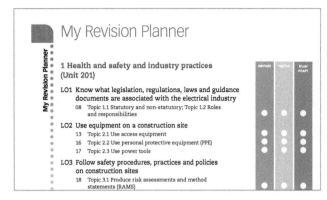

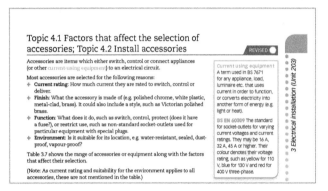

Features to help you succeed

Exam tips

Expert tips are given throughout the book to help you polish your exam technique in order to maximise your chances in the exam.

Typical mistakes

The author identifies the typical mistakes that candidates make in exams and explains how you can avoid them.

Now test yourself

These short, knowledge-based questions provide the first step in testing your learning. Answers are available online.

Definitions and key words

Clear, concise definitions of essential key terms are provided where they first appear.

Exam checklist

The exam checklists provide a quick-check bullet list for each topic.

Exam-style questions

Practice exam questions are provided for each topic. Use them to consolidate your revision and practise your exam skills.

Online

Go online to check your answers to the exam questions at **www.hoddereducation.co.uk/myrevisionnotesdownloads**

Check your understanding

These questions test your basic understanding of the information as you work through the course. Answers are available online.

Exam breakdown

For guidance on how you will be assessed and how to prepare for your exam, see the end of this book (page 117).

My Revision Notes: City & Guilds Level 2 Advanced Technical Diploma in Electrical Installation (8202-20)

My Revision Planner

REVISED   TESTED   EXAM READY

Check your understanding and progress at **www.hoddereducation.co.uk/myrevisionnotes**

REVISED TESTED EXAM READY

My Revision Planner

	REVISED	TESTED	EXAM READY

Check your understanding and progress at **www.hoddereducation.co.uk/myrevisionnotes**

Countdown to my exams

6–8 weeks to go

+ Start by looking at the specification — make sure you know exactly what material you need to revise and the style of the examination. Use the revision planner on pages 4–6 to familiarise yourself with the topics.
+ Organise your notes, making sure you have covered everything on the specification. The revision planner will help you to group your notes into topics.
+ Work out a realistic revision plan that will allow you time for relaxation. Set aside days and times for all the subjects that you need to study and stick to your timetable.
+ Set yourself sensible targets. Break your revision down into focused sessions of around 40 minutes, divided by breaks. These Revision Notes organise the basic facts into short, memorable sections to make revising easier.

REVISED ○

2–6 weeks to go

+ Read through the relevant sections of this book and refer to the exam tips, summaries, typical mistakes and key terms. Tick off the topics as you feel confident about them. Highlight those topics you find difficult and look at them again in detail.
+ Test your understanding of each topic by working through the 'Now test yourself' questions in the book. Look up the answers online.
+ Make a note of any problem areas as you revise, and ask your teacher to go over these in class.
+ Look at past papers. They are one of the best ways to revise and practise your exam skills. Write or prepare planned answers to the exam practice questions provided in this book. Check your answers online at **www.hoddereducation.co.uk/ myrevisionnotesdownloads**
+ Track your progress using the revision planner and give yourself a reward when you have achieved your target.

REVISED ○

One week to go

+ Try to fit in at least one more timed practice of an entire past paper and seek feedback from your teacher, comparing your work closely with the mark scheme.
+ Check the revision planner to make sure you haven't missed out any topics. Brush up on any areas of difficulty by talking them over with a friend or getting help from your teacher.
+ Attend any revision classes put on by your teacher. Remember, he or she is an expert at preparing people for examinations.

REVISED

The day before the examination

+ Flick through these Revision Notes for useful reminders, for example, the exam tips, typical mistakes and key terms.
+ Check the time and place of your examination.
+ Make sure you have everything you need – extra pens and pencils, tissues, a watch, bottled water, sweets.
+ Allow some time to relax and have an early night to ensure you are fresh and alert for the examinations.

REVISED

My exams

8202-20: Advanced Diploma in Electrical Installation

Date: ....

Time: ....

Location: ....

1 Health and safety and industry practices (Unit 201)

Each year, many deaths and thousands of injuries occur in the workplace, with a large proportion occurring in the construction and building service industries. A positive approach to health and safety legislation and an understanding of the hazards, risks and risk reduction methods will go a long way to reducing those figures.

This chapter revisits the legislation, procedures and practices that will help you in your future career, as well as your forthcoming exam. We will also take another look at environmental protection requirements, practices, and the structure and roles within the construction industry.

> **Hazard** Something that is dangerous and could cause harm (e.g. working at height).
>
> **Risk** How likely a hazard is to cause harm and how much harm it could cause.

LO1 Know what legislation, regulations, laws and guidance documents are associated with the electrical industry

The construction industry is regulated by many statutory and non-statutory documents. While you do not need to know them all inside out, you do need to know they are there.

Topic 1.1 Statutory and non-statutory; Topic 1.2 Roles and responsibilities

REVISED

There are many statutory and non-statutory regulations, guides and laws that control or regulate site-based activities. The tables below show each of these documents, their legal status, what they regulate and who they apply to. Many of the statutory regulations are maintained and enforced by the Health and Safety Executive (HSE).

Tables 1.1 to 1.16 show the statutory legislation and regulations that you must know.

> **Statutory** The regulations are law and must be followed.
>
> **Non-statutory** Not law but following them is considered as best practice.
>
> **Health and Safety Executive (HSE)** The UK body responsible for shaping and reviewing health- and safety-related regulations, producing research and statistics, and enforcing the law.

Table 1.1

The Health and Safety at Work etc. Act	
Legal status and who maintains it:	Statutory UK parliament and enforced by HSE
What does it cover?	+ It is known as an enabling act – it gives powers to the HSE to produce detailed regulations that are specific to work-related tasks. + It provides general legislation, covering occupational health and safety and voluntary work (hence the 'etc.' in the title).
Who does it apply to?	It sets out the general duties that: + employers have towards employees and others such as members of the public + employees have to themselves, as well as co-workers + it also applies to self-employed persons in the same way.

> **Exam tip**
>
> The Health and Safety at Work etc. Act can be abbreviated in exams to HSWA or HASWA.

Table 1.2

The Electricity at Work Regulations	
Legal status and who maintains it:	Statutory HSE
What does it cover?	Most of the regulations are directed at building electrical installations. Installations must: + be of proper construction + have conductors which are properly insulated (or other precautions taken) + have a means of cutting off the power for electrical isolation. There are also some regulations stating principles of safe working practice (e.g. Regulation 14, which covers live working, is very important). There are also regulations specific to hazardous locations such as mines.
Who does it apply to?	+ All persons working on or near electrical systems (anyone at work). + A building or facility must appoint a duty holder who is responsible for ensuring electrical safety is maintained. + Persons working on electrical systems have a duty to protect themselves and others.

Table 1.3

The Management of Health and Safety at Work Regulations	
Legal status and who maintains it:	Statutory HSE
What does it cover?	It requires employers to: + carry out risk assessments + then to make arrangements to implement safety measures based on the assessment + appoint competent people + arrange for appropriate information and training relating to safety
Who does it apply to?	Employers and self-employed persons.

Table 1.4

Workplace (Health, Safety and Welfare) Regulations	
Legal status and who maintains it:	Statutory HSE
What does it cover?	A wide range of basic health, safety and welfare issues such as ventilation, heating, lighting, workstations, seating and welfare facilities.
Who does it apply to?	+ Employers who provide facilities for employees. + Includes a section relating to temporary work sites.

Table 1.5

Control of Substances Hazardous to Health (COSHH) Regulations	
Legal status and who maintains it:	Statutory HSE
What does it cover?	Persons must assess the risks from hazardous substances and take appropriate precautions.
Who does it apply to?	+ Employers and self-employed have a duty to make the assessment. + Employees must follow precautions.

Check your understanding

1 What is the document number that is the Approved Code of Practice for the COSHH regulations? (You can have a look on the HSE website.)

My Revision Notes: City & Guilds Level 2 Advanced Technical Diploma in Electrical Installation (8202-20)

Table 1.6

Working at Height Regulations	
Legal status and who maintains it:	Statutory HSE
What does it cover?	✛ Preventing death and injury caused by a fall from height. ✛ It also includes requirements on the maintenance and use of access equipment (e.g. ladders and scaffolding).
Who does it apply to?	✛ Employers and those in control of any work at height activity must make sure work is properly planned, supervised, and carried out by competent people. ✛ Employees have general legal duties to take care of themselves and others who may be affected by their actions.

Table 1.7

Personal Protective Equipment (PPE) at Work Regulations	
Legal status and who maintains it:	Statutory HSE
What does it cover?	PPE should be used as a last resort. Wherever there are risks to health and safety that cannot be controlled in other ways, the Regulations require PPE to be supplied. The Regulations also require that PPE is: ✛ properly assessed before use to make sure it is fit-for-purpose ✛ maintained and stored properly ✛ provided with instructions on how to use it safely ✛ used correctly by employees.
Who does it apply to?	✛ Employers should provide PPE and use it correctly. ✛ Self-employed persons.

> **Exam tip**
>
> Always remember when choosing methods of risk reduction, PPE is a last resort.

Table 1.8

Manual Handling Operations Regulations	
Legal status and who maintains it:	Statutory HSE
What does it cover?	These regulations apply to any transporting or supporting of a load (e.g. lifting, putting down, pushing, pulling, carrying or moving by hand or bodily force).
Who does it apply to?	All persons involved in the activity of manual handling – but employers should provide suitable equipment.

Table 1.9

Provision and Use of Work Equipment Regulations	
Legal status and who maintains it:	Statutory HSE
What does it cover?	These regulations place responsibilities on businesses and organisations whose employees use work equipment – it must be safe, suitable and persons must be suitably trained to use it.
Who does it apply to?	Employers and self-employed persons.

Check your understanding and progress at **www.hoddereducation.co.uk/myrevisionnotes**

Table 1.10

Control of Asbestos at Work Regulations	
Legal status and who maintains it:	Statutory HSE
What does it cover?	+ Gives minimum standards for protecting employees from risks associated with exposure to asbestos. + Has requirements for certain types of non-licensable work with asbestos, notification of work, designating areas where you are working on asbestos, medical checking and record keeping (such as an asbestos register).
Who does it apply to?	+ Building owners or occupiers and those working on materials containing asbestos.

Table 1.11

Environmental Protection Act	
Legal status and who maintains it:	Statutory Secretary of State for Environment, Food and Rural Affairs/ Environment Agency
What does it cover?	Gives the minimum requirements and responsibilities for waste management and control of emissions into the environment.
Who does it apply to?	Anyone undertaking waste disposal or emitting polluting substances into the air, land or water.

Table 1.12

The Hazardous Waste Regulations	
Legal status and who maintains it:	Statutory Secretary of State for Environment, Food and Rural Affairs/ local authorities
What does it cover?	Places a 'duty of care' on businesses to ensure hazardous waste produced, handled or transported causes no harm.
Who does it apply to?	Employers and self-employed

Table 1.13

Pollution Prevention and Control Act	
Legal status and who maintains it:	Statutory Secretary of State for Environment, Food and Rural Affairs/ local authorities/the Environment Agency
What does it cover?	+ Covers installations that emit pollution. + Places a duty on the operator to eliminate or reduce the pollution to harmless levels. + These installations must be licensed.
Who does it apply to?	Installation operators

Table 1.14

Control of Pollution Act	
Legal status and who maintains it:	Statutory The Environment Agency
What does it cover?	Allows powers to the Environment Agency to produce regulations on waste disposal, water pollution, noise and atmospheric pollution, as well as public health.
Who does it apply to?	Anyone involved in acts involving any pollution.

Table 1.15

The Control of Noise at Work Regulations	
Legal status and who maintains it:	Statutory HSE
What does it cover?	Ensure that workers' hearing is protected from excessive noise at their place of work. Aims to avoid workers losing their hearing and/or suffering from tinnitus (permanent ringing in the ears).
Who does it apply to?	Anyone at a place of work.

Check your understanding

2 What two items of PPE can protect your ears from loud noise while at work?

Table 1.16

The Waste Electrical and Electronic Equipment (WEEE) Regulations	
Legal status and who maintains it:	Statutory The Environment Agency
What does it cover?	+ Gives requirements for the recovery, reuse, recycling and treatment of most forms of electrical and electronic waste (any equipment that has a plug, power supply or battery). + All manufacturers and distributors of electrical products to establish an infrastructure where all households have a facility for returning those goods. + Figure 1.1 shows the WEEE symbol, which is displayed on products covered by the Regulations.
Who does it apply to?	Anyone disposing of a WEEE-covered product should take it to a suitable recycling centre.

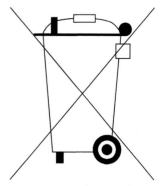

Figure 1.1 The WEEE symbol found on products covered by the Regulations

Tables 1.17 to 1.20 cover the non-statutory legislation and regulations that you must know.

Table 1.17

BS 7671: Requirements for Electrical Installations	
Legal status and who maintains it:	Non-statutory The Institute of Engineering and Technology (IET)/British Standards Institute (BSI)
What does it cover?	+ Technical requirements for all electrical installation work in the UK (but not those covered by specific legislation, e.g. mines). + Covers design, installation and maintenance (periodical inspections) of electrical systems in buildings and structures up to 100 V AC and 1500 V DC – including the erection of data and signal cables.
Who does it apply to?	Anyone installing electrical systems in or on buildings, structures or facilities.

Check your understanding

3 Which IET guidance publication are you allowed to take into your Level 2 exam?

Table 1.18

IET Guidance	
Legal status and who maintains it:	Non-statutory IET
What does it cover?	There are eight Guidance Notes (GN) – these detail specific parts of BS 7671, e.g. earthing (GN8), selection and erection (GN1), or inspection and testing (GN3).
Who does it apply to?	Anyone installing electrical systems in or on buildings, structures or facilities.

Table 1.19

HSE Guidance Publications	
Legal status and who maintains it:	Non-statutory HSE
What does it cover?	✢ The HSE produces many guidance documents and publications to simplify statutory regulations. ✢ The documents are coded depending on what they cover (e.g. INDG documents are industrial guidance – posters or charts providing easy-to-follow guidance). ✢ Health and safety guidance (HSG) documents give health and safety guidance on specific situations (e.g. Avoiding Danger from Underground Services (HSG47)). ✢ Legal guides (L documents) contain statutory regulations and guidance on how to comply with those legal duties (e.g. L25 covers PPE regulations).
Who does it apply to?	Anyone who requires guidance on how to comply with statutory duties.

Table 1.20

Approved Codes of Practice (ACOP) and Codes of Practice (CoP)	
Legal status and who maintains it:	Non-statutory, but following the guidance will lead to compliance with statutory documents. CoP are maintained by either the HSE or institutes responsible for sectors of industry such as the IET.
What does it cover?	✢ The HSE produce a wide variety of ACOPS. ✢ Many industry bodies (e.g. the IET) produce codes of practice (CoP), such as the IET CoP for the in-service inspection and testing of electrical equipment. It states how to comply with the Electricity at Work Regulations in relation to electrical appliances.
Who does it apply to?	Anyone who requires guidance on how to comply with statutory duties.

LO2 Use equipment on a construction site

Many of the accidents that occur on construction sites, or while carrying out work activities, happen when specialist equipment is being used, such as access equipment or power tools.

Topic 2.1 Use access equipment

The Working at Height Regulations cover the use of access equipment. HSE guidance, such as INDG455, provides easy-to-follow guidance on how to safely use particular types of access equipment.

Your exam could have questions relating to four items of access equipment. These are:

✚ steps or stepladders
✚ ladders
✚ mobile scaffold towers
✚ platforms.

Here is some detail for each item relating to pre-use checks, suitability, erection and use.

> **INDG455** An industry guidance document published by HSE on the safe use of ladders and stepladders. INDG guides are free to view on the HSE website at https://www.hse.gov.uk

With all access equipment, if any task is being undertaken while using the access equipment, the equipment should be positioned to prevent over-reaching and the user should only ever carry lightweight materials or tools.

Pre-user checks: Steps, stepladders and ladders

+ Check the stiles:
    + Make sure they are not damaged, as the ladder could collapse.
+ Check the feet:
    + If they are missing, worn or damaged, the ladder could slip. Also check ladder feet when moving from soft or dirty ground. Debris on the feet could cause the ladder to slip on a smooth surface.
+ Check the rungs or treads:
    + If they are bent, worn, missing or loose, the ladder could fail, causing a fall from height. Check for contamination making them slippery.
+ Check any locking mechanisms:
    + Check they are not bent, or if any fixings are worn or damaged, as the ladder could buckle and collapse. Ensure any locking bars are fully engaged.
+ Check the stepladder platform:
    + This acts as part of the locking mechanism – if it is split or buckled, the ladder could become unstable or collapse.

Some organisations use ladder tags attached to the steps or ladders. These tags show an inspection has been undertaken. If it has, you must still check it before use.

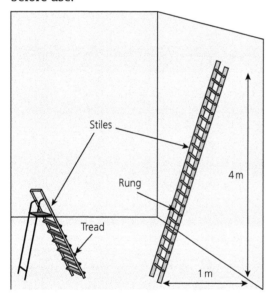

Figure 1.2 Using ladders and stepladders

Using stepladders

+ Check all four feet are in contact with the ground and the ground is level.
+ Do not stand on the top three steps (including the top step) unless there is a suitable handhold.
+ Try to position the stepladder so you are facing what you are working on. If this isn't possible due to space, and you work side on, tie the steps to secure them from tipping.
+ Always maintain a minimum three points of contact with the stepladder. If using two hands to carry out a task, two feet and the body should be supported by the ladder but this stance should be adopted only for short periods.

Check your understanding and progress at **www.hoddereducation.co.uk/myrevisionnotes**

Setting up ladders

When setting up ladders, the following should be observed:

+ Ladders should never be rested against soft materials such as guttering. Use a stand-off where needed.
+ Never extend a ladder while standing on a rung.
+ Always erect a ladder one unit from the wall for every four units up.
+ Only stand a ladder on firm ground.
+ Always secure a ladder by tying at the top, or the bottom, or by a wall tie near the base. If it cannot be secured, or during securing at the top, the ladder should be footed by another person.
+ Where no alternative exists, ladders may be tied part way down around a suitable structural support such as an open window.
+ Ladders used to access platforms should always extend 1 m above the landing point to give suitable handhold.

> **Stand-off** A sturdy attachment to a ladder that enables the ladder to rest standing off from a wall around 0.5 m, leaving the top of the ladder clear from resting against brittle surfaces such as gutters or windows.

Now test yourself TESTED

1   A ladder needs to reach a vertical distance of 4 m to a landing platform and maintain the correct ratio and handhold requirements. How long should the ladder be?
2   Which HSE guidance document details safe and practical use of ladders?

> **Exam tip**
>
> Exam questions may need you to apply trigonometry when working out ladder lengths. You can revise this in Chapter 2 of this book.

When using ladders for carrying out a task

+ Maintain three points of contact with the ladder at all times.
+ Always face the ladder while climbing and descending.
+ Avoid carrying items; try to use a tool-belt.
+ Do not work within 6 m of any powerline unless it has been made dead.

Erecting and using mobile scaffold towers and other platforms

Mobile scaffold towers and platforms are used to perform tasks at height where a ladder or steps are unsuitable. This could be due to the duration of the task or because heavier objects are needed, such as tools or materials.

Check your understanding

4   What is an MEWP?

> **Platforms** Defined by the Working at Height Regulations as any surface above ground height used as a place of work or as means of access to a place of work and includes scaffold, suspended scaffold, cradles and mobile platforms.

Rules regarding mobile scaffold towers and work platforms include the following:

+ Tower scaffolding and other platforms over 2 m in height from the ground should have a documented inspection carried out on it following installation.
+ Always erect towers and platforms using the instructions.
+ Never rest ladders or other access equipment onto a tower or platform.
+ Never use a tower in strong winds.
+ Never move a tower which is 4 m or more in height.
+ Always check for obstructions or potholes on the route before moving equipment.

Check your understanding

5   Research the HSE website for information on erecting and using tower scaffolding. Who has a duty to provide instructions for erecting the tower?

Topic 2.2 Use personal protective equipment (PPE)

REVISED

Table 1.21 shows the most common types of PPE, their purpose and any specific variations of each type.

Table 1.21

General item	Purpose	Variants
Footwear	+ To protect toes from falling objects or being crushed. + To protect the underside of the foot from hazardous objects (e.g. nails). + To protect other parts of feet and ankles from abrasion, oils or chemicals. + To keep feet dry. + To provide good grip.	+ Steel-toe capped or hardened boots/shoes/trainers + Wellingtons (wellies) + Rigger boots
Pads	+ To protect knees when frequently kneeling down to carry out tasks. + Elbow pads are sometimes used to protect elbows in tight spaces from bashing against nearby surfaces.	+ Separate pads that attach to knee + Incorporated into trousers + Kneeling cushions
Harness	+ To limit the risk of falling from a height by attaching to a fixed anchor by a fall-arrest lanyard.	+ Belt harness + Full body harness
Suits	+ To protect the body or clothing from getting wet or damage by oils/chemicals/splashes.	+ Cloth coverall (boiler suit) + Disposable coveralls + Bib and brace + Waterproofs

Gloves	+ To protect hands against:
	  + abrasion
	  + cuts
	  + burns
	  + chemicals
	  + contamination
	+ To provide good grip when using tools etc.
		+ Gloves
		+ Mitts
		+ Gauntlets
		+ Rigger

High visibility clothing (hi-vis)	+ Ensures a person can be easily seen by others who may be operating moving machinery, such as cranes or vehicles, to avoid accidents.	+ Vests + Jackets + Trousers + Sashes + Body-warmers

Eyewear	+ To protect eyes from:
	  + contamination by dust, dirt, swarf, splashes, sprays, etc.
	  + arcing, such as welding.
		+ Glasses
		+ Goggles
		+ Visors

Respiratory	+ To prevent or reduce breathing in contaminants or fumes or provide an oxygen supply.	+ Dust masks + Respirators + Powered air units
Ear wear	+ To prevent or reduce hearing damage from noise.	+ Ear plugs + Ear defenders
Headgear	+ To protect the head from falling objects or above head/head height hazards.	+ Hard hats + Bump caps

Check your understanding

6 Research the internet to see which HSE guidance document provides key information relating to PPE.

Exam tip

Do not confuse PPE with items used to protect the fabric of the building such as boot covers. They are there to protect the carpets and floors – not you.

When selecting and using PPE, remember:

+ PPE is considered the last line of defence. All other risk reduction measures must be taken before PPE is used as a means of protection.
+ PPE may increase risk. It should be assessed for suitability (e.g. dust masks may stop someone from communicating properly).
+ Items must be compatible (e.g. will a dust mask make eye protection difficult to fit correctly?).

Check your understanding

7   Why should gloves be avoided when operating machinery such as a bench drill?

Topic 2.3 Use power tools

REVISED ○

Power tools are risky items that can cause harm, especially when used on construction sites. This is because it can be a hazardous environment, which increases the risk of damage or deterioration due to:

+ mechanical damage to casings, extension leads or flexible cables
+ chemical damage to leads and cables
+ failure of connections and cable grips due to stretching and pulling cables
+ poor functioning of machinery due to contamination or poor maintenance.

One major risk with using power tools that have become damaged is electric shock as the casings and cable insulation provide basic protection against electric shock.

Table 1.22 shows the risk from power tools is also dependent on other factors.

Table 1.22

Voltage	+ 110 V power tools are preferred to 230 V versions. + If suitable, battery-operated tools are safer.
Rating	+ Higher-rated appliances (in watts) can create a higher risk of injury. + For example, drill snatch – when a drill bit gets stuck, causing the drill body to spin. Users have more chance of controlling lower-rated tools. + On average, users have more chance of controlling tools up to 300 W.
Class	There are three common classes of power tool. They apply to tools that are connected to an external power source, such as a socket-outlet: + **Class I:** + Have earthed metal parts that could become live if a fault occurred. + It is essential the earth connection remains well connected to protect the user from the risk of shock by causing the circuit to disconnect. + **Class II:** + These parts have reinforced insulation, protecting the user from the risk of shock. + They are preferred to Class I items – but this is not always possible if the tool has multiple metallic parts. + **Class III:** + These are items that have a suitable extra-low voltage supply, such as a 25 V hand-lamp. Battery-operated devices will fall under the class system if plugged in to charge. This is usually Class II or III.

Mechanical damage
Damage such as tears, cuts, abrasion, crushing (or similar).

Basic protection A technical term used in BS 7671 – refers to the insulation around live parts, or the barriers and enclosures housing live parts, which prevent users from touching live parts.

Earth Also known as 'terrafirma' from the French word for 'earth' – 'terre'. Many international electrical regulations are written in French, so 'T' is the symbol for 'earth' when looking at earthing arrangements.

Exam tip

Remember the plug and socket voltage colours.

Check your understanding

8   What do the symbols in Figure 1.3 indicate?

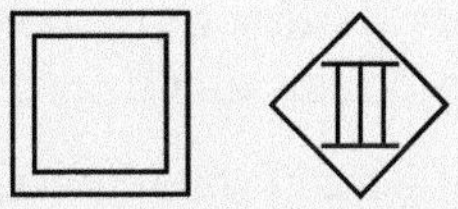

Figure 1.3

17

User checks

Before using power tools (or any electrical appliance) a user check should be carried out. This includes checking the following items, as listed in Table 1.23 below.

User check A term used in the IET Code of Practice for in-service inspection and testing of electrical equipment (CoP ISITEE), which is considered a vital safety precaution before using any electrical equipment.

Table 1.23

Plugs	+ Check for signs of damage, such as bent pins; case has no cracks or damage. + No signs of overheating. + Cable grips are secure, and the cable doesn't move. + Plug does not rattle, which indicates loose internal parts.
Flexible cable	+ In good condition. + Free from cuts, grazing or damage. + Not located in a position where it could be damaged. + No signs of temporary repairs such as tape on joints. + Not too tightly bent. + Not a trip hazard. + If using an extension lead, is it fully uncoiled to avoid overheating.
Socket-outlet	+ No sign of overheating or damage.
RCD on lead or socket	+ Check the RCD function by pressing the test button, then re-set.
Equipment casing	+ Free from cracks, chemical corrosion or damage. + Flexible cable secure as it enters casing and not pulled or stretched. + No signs of overheating.
Suitability	+ Equipment is suitable for the environment and the intended use.

Always remember that between tasks, power tools may become damaged because of other site activities or poor storage. **Always** consider a user check before using.

LO3 Follow safety procedures, practices and policies on construction sites

Construction sites are very hazardous places. Following specified policies and procedures is essential for the safety of everyone. Always remember: your actions can put you and others at risk.

Topic 3.1 Produce risk assessments and method statements (RAMS)

REVISED

A risk assessment is a fairly simple task which most people do on a day-to-day basis. The only difference between a day-to-day assessment and a *documented* risk assessment is writing it down to prove it has been made, and for you and others to refer to it.

Method statements are documented instructions on *how* to carry out a task to comply with the risk assessment and be as safe as possible.

For example: when you need to cross a busy road, which is a hazard, you check the risk of being hit by a vehicle before stepping into the road. You stop, look for vehicles in both directions and listen for vehicles you may not be able to see.

Method A set way to do something; a method statement is the method in writing.

This is effectively a risk assessment to see how likely you are to be hit, decide on how safe it is to cross, or to look for alternative safer methods of crossing the road, such as a bridge or pedestrian crossing.

If the road is very busy but a suitable crossing is nearby, you follow a set method to use the crossing:

+ Stop on the pavement.
+ Press the crossing button.
+ Wait for the green crossing light to indicate it may be safe to cross.
+ Check that all traffic has stopped and the crossing is clear.
+ Cross to the other side while continuing to scan the traffic.
+ Step onto the pavement on the other side.

If this procedure was documented, it would be a method statement.

Five steps to completing a risk assessment

The HSE has created a five-step approach to completing a risk assessment. This is shown in Table 1.24.

Table 1.24

Step	Action
1 Identify the hazard	What are you about to do: use a power drill, work at height, or carry a heavy object? Nearly every single task is a potential hazard.
2 Who might be harmed	Think about who may be harmed, and how, by the hazard. This gives a clearer picture of how the risk can be controlled.
3 Evaluate the risk	+ What is the level of risk? + How likely is it to occur? + How can it be removed or reduced? Consider methods of risk reduction, such as: + trying a less risky option + preventing access by others to the hazardous area + consulting others + working timings around others. PPE should always be the last resort as a method of risk reduction.
4 Record your findings	Record all the above on a risk assessment document. Keep it as simple as possible, focusing on control methods. Written records are a good way to reassess so once the risk has been recorded, look at it again – ensure there is nothing missed and that the risk is low after measures are in place.
5 Regularly review	Few sites remain the same. Changes in location will change the level of risk, so always review a completed risk assessment to ensure the following: + There are no major changes. + Can you make further improvements? + What did you learn from the last time the assessment was made? Did something happen that wasn't thought of last time?

Now test yourself   TESTED ◯

3   Go to the HSE website and download a risk assessment template (free to use). Complete the risk assessment based on a simple everyday task, such as making toast.

Young person's risk assessment

Risk assessments should assess the risk for all involved in a task, whether they are actively doing the task or are close by to it.

However, young persons can introduce additional risks because of inexperience of the workplace, being unaware of the risks and a potential lack of maturity.

> **Young person** Someone aged between 16 and 18, but may include 15 if they turn 16 in that academic year. A young person under the age of 16 is considered a child.

As a result, employers must consider further factors that impact on the risk that may be a challenge for a young person, such as:

+ workplace layout
+ any hazardous materials
+ handling work equipment
+ methods or processes.

Topic 3.2 Types of accident reporting

REVISED

Even when all risk reduction methods have been put into place, accidents still happen!

Accidents can be minor, needing little or no attention. But they could be major or cause long-term effects. The severity of the accident will affect the further actions required.

RIDDOR

Under the Reporting of Injuries, Diseases and Dangerous Occurrences Regulations (RIDDOR), certain workplace accident situations need to be reported to the HSE.

Specified injuries or situations include:

+ fractures (other than to toes, thumbs and fingers)
+ amputations
+ loss or partial loss of sight
+ crush to brain or internal organs
+ serious burns
+ loss of consciousness caused by head injury or breathing difficulty
+ injuries arising from working in confined location
+ injuries resulting in absence from work for over seven days
+ accidents to non-workers, such as members of the public
+ occupational diseases
+ dangerous occurrences and specified near-misses, such as scaffold collapse.

Note: if somebody is off work for more than three days, but less than seven, this must be recorded – but does not need reporting to HSE. This could be reported in an accident book. Other injuries which result in seven or more days off work require reporting to the HSE.

> **Check your understanding**
>
> 9 How soon after an accident should the following incidents be reported to the HSE?
>   a) An accident resulting in death.
>   b) An accident resulting in bad bruising, but the person was off work for nine days.

Reporting accidents to the emergency services

When accidents or incidents occur at work, an immediate reaction is to ring the emergency services for help. When you dial 999 or 112, the person on the other end of the phone will need some information. Before you call, think about the following:

+ Address and location of the incident.
+ Nature of the incident (e.g. fire, accident).
+ Any difficulties that the emergency services may have getting to the incident.
+ Any immediate dangers, such as leaking gas or fuel, or persons trapped.

If you need to contact the emergency services, make sure you are clear and concise with your information.

> **Check your understanding**
>
> 10 What emergency services could you contact by calling 999 or 112?

Topic 3.8 Site safety management

REVISED ⬤

A clear emergency action plan on a site makes sure that everyone is aware of what to do if an incident happens. It can reduce panic and confusion as everyone understands their role.

On construction sites, this information is often given during the safety induction for the site. A good emergency plan would consider the following:
+ Name of appointed person who takes control of any incident.
+ Name of appointed person responsible for first aid.
+ Locations of first aid facilities.
+ Who is responsible for the facilities (keeping them stocked)?
+ What should happen in the event of an emergency or incident?
+ What should happen after this incident?

Always remember that some buildings and facilities have different evacuation procedures to others due to the risk.

> **Check your understanding**
>
> **11** What is meant by vertical and horizontal evacuation plans?

The person appointed as the responsible person on a site will ensure the following:
+ Everyone is aware of the common site rules and when conditions change, this is communicated by toolbox talks (or similar).
+ Orientation is assessed and monitored. This includes where facilities are and arranging one-way systems where pinch points occur in tight spaces.
+ Prohibition areas are monitored. These are areas where restricted or no access applies due to a risk or a specific operation, such as laying a wet concrete floor.
+ Access and egress routes are kept free to allow safe entry, exit and emergency routes out.

> **Toolbox talks** Regular, on-site, informal safety meetings. Some sites have them daily before work starts.

Topic 3.3 Principles of fire safety

REVISED ⬤

A fire needs three ingredients to start and spread. We remember this by using the fire triangle.

Fires are categorised into different classes, depending on the fuel source or cause. It is vitally important that the correct coded fire extinguisher is used on the correct class of fire.

Figure 1.5 below shows the different classes of fire and, as a guide, indicates the correct extinguisher to use.

Figure 1.4 The fire triangle

Fire classification	Water	Foam	Powder	CO$_2$
Class A – Combustible materials such as paper, wood, cardboard and most plastics	✓	✓	✓	
Class B – Flammable or combustible liquids such as petrol, kerosene, paraffin, grease and oil		✓	✓	✓
Class C – Flammable gases, such as propane, butane and methane			✓	✓
Class D – Combustible metals, such as magnesium, titanium, potassium and sodium			Specialist dry powder	
Class E – electrical fires				✓
Class F – Cooking oils and fats		Specialist wet chemical (yellow)		

Figure 1.5 A guide to the different classes of fire and the types of extinguisher that could be used

21

TESTED ◯

Now test yourself

4 When a fire starts, people might easily panic and try to tackle a fire using the wrong fire extinguisher. What would happen if the following types of fire were tackled using a fire extinguisher with a red band?
   a) Fire caused by electrical equipment.
   b) A chip fryer full of oil.

What to do in the event of a fire

If a fire starts, the correct procedure to save as many lives as possible is as follows:

+ Raise the alarm immediately to give everyone the best chance to escape. Contact the emergency services.
+ If trained and safe to do so, tackle the fire using equipment provided.
+ If firefighting fails, leave the building immediately.
+ Ensure everyone has left the building or site.
+ Close as many doors as possible to contain the fire.
+ Go to the designated assembly point which has been chosen to keep persons away from any firefighters arriving.

Topic 3.4 Manual handling principles

REVISED ◯

Manual handling of objects is one of the major causes of work-based injuries. Poor posture and poorly considered routes can make a simple task hazardous.

Before manually moving a load, consider:
+ Is the route to be taken clear with no hazards?
+ Can the load be moved by other mechanical aids?
+ How will the load be gripped and where is the most weight?

> **Mechanical aids** Items used to make handling materials easier and safer, such as trolleys, sack barrows and dollies.

When lifting the load:
+ Bend your knees, keeping a straight back.
+ Grip the load at the base or where directed.
+ Bring the load as close as possible into your body.
+ Stand up by straightening your knees.
+ Bend your knees, not your back, to place the load down.

Topic 3.5 Types of signage used on a construction site

REVISED ◯

There are **four** groups of signs used on construction sites.

Prohibition signs

These signs indicate an activity that must not be done. They are circular white signs with a red border and red cross bar.

No access for unauthorised persons

No smoking in this building

In the event of fire do not use this lift

Do not drink

Figure 1.6 Various prohibition signs

Check your understanding and progress at **www.hoddereducation.co.uk/myrevisionnotes**

Warning signs

These provide safety information and/or give warning of a hazard or danger. They are triangular yellow signs with a black border and symbol.

Danger
High voltage

Warning
Trip hazard

Caution
Work in progress

Warning
Falling objects

Figure 1.7 Warning signs

Mandatory signs

These signs give instructions that must be obeyed. They are circular blue signs with a white symbol.

Hand protection
must be worn
in this area

Eye protection
must be worn

High visibility clothing
must be worn in
this area

Safety helmets
must be worn
in this area

Figure 1.8 Mandatory signs

Advisory or safe condition signs

These give information about safety provision. They are square or rectangular signs with a white symbol.

Fire
assembly
point

First aid
box

Fire
exit

Figure 1.9 Advisory signs

Topic 3.6 Procedures for common hazardous materials

Electricians do not use many hazardous substances in the majority of their tasks but common ones they do deal with include:

+ adhesives: such as those used to weld PVC conduits and accessories
+ resins: such as those used in underground cable joints
+ acids: such as lead acid in batteries
+ fuels/oils: used in generators or other engines.

When dealing with substances that are covered by the Control of Substances Hazardous to Health (COSHH) Regulations, several things must be considered:

+ **Storage:** Does the substance need to be stored in cool or refrigerated areas, or well-ventilated areas?
+ **Use:** For example, adhesives should not be used in confined locations; locations should be well-ventilated.
+ **Disposal:** Is the packaging likely to contain residual amounts of the substance that could be a hazard to persons processing the waste? If so, it should be disposed of at a special facility.

The manufacturers' data sheet provides detailed instructions for how to store, use and dispose of the substance safely.

Employers and the self-employed have a duty to risk-assess substances under COSHH requirements and implement suitable control measures.

> **Exam tip**
>
> Remember that all risk reduction measures need to be in place before PPE is considered. So, good ventilation comes before breathing aids.

Packaging signs

Hazardous substances will include CHIPS labels on the packaging, indicating the type of substance.

> **CHIPS** Stands for Chemical Hazard Information and Packaging for Supply Regulations.

Figure 1.10 The hazard signs shown indicate substances that are (top row, left to right) corrosive, dangerous to the environment, gas under pressure, (middle row, left to right) requires caution, toxic, long-term health hazards (causes of cancer), (bottom row, left to right) oxidising, explosive, flammable

Topic 3.7 Procedures for dealing with asbestos

Asbestos was widely used as a building material from the 1950s up to the 1980s. It can be found in many parts of a building. It is generally safe unless disturbed. When disturbed, its fibres are a very high risk to long-term health by causing lung disorders.

Every place of work, including construction sites on existing buildings or demolition sites, must have an asbestos register. This details where asbestos is known or likely to be.

Work on or around asbestos requires full risk assessments and a method statement detailing control measures, including air monitoring and PPE requirements.

If asbestos is encountered where not known or suspected, the person responsible for the building or site must be informed immediately.

Asbestos can only be removed from a site by a specialist contractor trained to deal with asbestos and licensed to dispose of it.

Common locations for asbestos are:
+ in rewireable fuse carriers
+ in ceilings that look like plaster ceilings
+ pipe lagging
+ as a heat insulator within boilers
+ roof and wall sheeting
+ as a fireboard around steel structures.

LO4 Carry out electrical safety procedures and practices

Working on electrical systems is a high-risk activity, even when the systems are switched off, as someone else could switch them back on again. Knowing the safe isolation procedure is one of the most essential skills of any electrician.

Safe electrical provisions on a construction site also contribute towards a safe site. Temporary supplies used by construction trades are essential to get the work done.

(Please note that Topic 4.3 is a practical performance outcome, so it will not be covered here.)

Topic 4.1 Safe isolation procedure

Table 1.25 outlines the procedures to be followed to safely isolate a circuit to work on. It includes what to consider for each stage of the process. This procedure must be followed to comply with the Electricity at Work Regulations. It explores all situations where isolation is required, such as isolating single-phase and three-phase:
+ installations
+ circuits, and
+ equipment.

Note: When electricians are working on an item of equipment, the circuit supplying that equipment should ideally be isolated, as many local devices cannot be secured in the off position. In these situations, a risk assessment should be undertaken to confirm the safest way to do this.

Table 1.25

Seek permission	+ Ensure it is safe to isolate the circuit. + Ensure the equipment on the circuit isn't essential for lighting or welfare (or it may create a hazard, such as supplies to lifts). + If the supply is for equipment that requires a shut-down procedure, such as computers or machinery, this must be followed before isolation. + Seek permission from the responsible person for the building or system.
Gather equipment	You will need: + An approved voltage indicator (AVI) to HSE GS38. + A proving unit or know the supply which checks the function of the voltage indicator. + A secure locking device and padlock with one unique key. If several people are working on the circuit, a multi-locking device is needed. + An 'Electrician Working' sign.
Identify	The point of isolation for the equipment or system; this could be: + The main switch for the entire installation. + The protective device for single circuit. + Local device for equipment, such as a fused connection unit or other devices which can be secured in the 'off' position.
Isolate	+ Turn the means of isolation off. + This is termed as placing the device in the 'open position' – opening the contacts cuts off the power.
Secure	+ Lock the isolating device in the open position. + Keep the key on your person so nobody else can access it. + Place the warning notice on the locking device.
Check	+ Check the condition of the AVI. + Check the function of the AVI on a known supply or proving unit.
Test	+ Test that the system or equipment is dead by testing, in the following sequence, on the load side of the isolating device/switch. **For single-phase systems or equipment:** **For three-phase systems or equipment:** Undertake the **'3-point test'** (count them) Undertake the **'10-point test'** (count them) L–N L1–N L2–N L3–N L–E L1–E L2–E L3–E N–E L1–L2 L1–L3 L2–L3 (L = Line, N = Neutral, E = Earth) N–E (L = Line, N = Neutral, E = Earth)
Prove	Once again, the AVI functions.
Confirm	Once again, check the securing device is secure.

Check your understanding

12 What does HSE document GS38 state regarding test leads used for electrical test equipment, such as the AVI used for safe isolation?

Exam tip

Remember the terms '10-point test' and '3-point test' to ensure you have covered all test points for isolation. Count them to be certain you have allowed for them all.

Check your understanding and progress at **www.hoddereducation.co.uk/myrevisionnotes**

Topic 4.2 Construction site supplies

Table 1.26 gives a summary of the types of electrical supplies provided on construction sites. These systems may be extensive for large construction sites or simple builder's board type power panels for small refurbishments.

Unless stated otherwise, many of the systems listed in Table 1.26 use 110 V centre tapped earth (CTE) supplies, which are provided by transformers having a 230 V input.

The 110 V supplies are known as 'reduced-low-voltage' systems, as the CTE arranges the 110 V into two phases, each having 55 V to earth, but the voltage difference between each phase is 110 V.

These are safer systems if faults happen, as the fault is only 55 V to earth.

> **Builder's board** A small but safe temporary supply. It usually consists of a wooden board near the incoming electrical supply. Mounted on the board is a distribution board fed from the incoming supply, with a range of socket-outlets to provide a power source while the building's systems are out of operation.

Table 1.26

Site lighting	Site temporary lighting can be in several forms: ✛ Festoon lighting is a string of lights on a cable used to illuminate walkways. ✛ Stand or task lights are lights on stands or tripods (more powerful than festoons) and used to illuminate work areas. ✛ Hand-lamps used for confined spaces – these are usually battery powered or powered at 25 V AC. ✛ Fixed lighting which is normally longer-term lighting on scaffolding. This may be supplied from a 230 V system with 30 mA RCD protection.
Site power	Most site power for tools and equipment is 110 V CTE provided by distribution centres. These house multiple 110 V (yellow) socket-outlets known as commando sockets. These socket-outlets conform to BS EN 60309-2. While 110 V is the most common voltage used, other power supply voltages may be present, and these socket-outlets are colour-coded to indicate their voltage. These are: ✛ Violet: 20–25 V AC ✛ White: 40–50 V AC ✛ Yellow: 110 V AC ✛ Blue: 200–250 V AC ✛ Red: 380–480 V AC
Charging stations	Most battery-powered equipment and tools come with chargers that have a standard 13 A plug, but 230 V standard socket-outlets are prohibited on most construction sites. As a result, many sites designate special charging areas or zones – 230 V sockets are provided for charging. The supplies are protected by 30 mA RCDs for additional protection.
Supply arrangements	Construction sites are dangerous environments with a high risk of impact damage and much contact with the ground. This increases the risk of electric shock so temporary electrical supplies need to be safe. Under the statutory requirements of the Electricity Safety, Quality and Continuity Regulations (ESQCR), electrical supply systems (known as PME) are **not** permitted for construction sites. As more electricity suppliers or distribution network operators (DNO) use PME more frequently, most supplies to construction sites need to be converted to TT systems. These are earthing arrangements that use an earth electrode as the means of earthing the site. BS 7671 section 704; Construction Sites, requires all electrical systems for construction sites to be protected by 30 mA RCDs (providing additional protection). As RCDs are very sensitive devices, 30 mA RCD protection will normally be provided for every temporary circuit, but a single time-delayed 30 mA RCD supplies the whole site. This means should a fault happen, the local RCD trips, leaving other circuits working. If the fault does not clear after a short time, the main RCD trips all circuits. Figure 1.11 shows a simple site supply system schematic diagram.

1 Health and safety and industry practices (Unit 201)

27

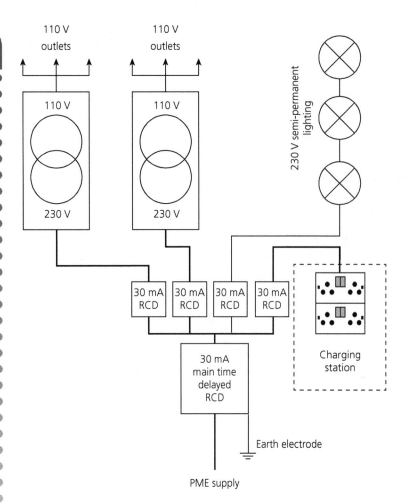

Figure 1.11 Simple site supply system schematic diagram

LO5 Understand environmental protection

The topics in this outcome look at contamination and waste disposal methods when dealing with products used for electrical activities. Many old products, such as fluorescent light fittings, contain hazardous substances and you may encounter them at work. If you do, you need to know what to do with them.

Topic 5.1 Types of waste management and disposal

REVISED

It wasn't very long ago when all waste, regardless of what it contained, went to landfill. This had to stop because, over time, chemicals can seep out of the waste. These chemicals contaminate the land and water and gases form, causing air pollution.

All waste must now be sorted and recycled for reuse where possible. As a result, waste products are categorised as shown in Table 1.27.

Table 1.27

Recycling	These are products that can be used again by re-processing, or re-purposed to form something else. These include: + **Cardboard and paper** – can be pulped and reused as cardboard or paper. + **Wood** – can be reused for a variety of purposes, e.g. shredding it to make furniture board such as MDF. + **Metal** – can be recycled by sawing and melting it down, or melting it and making products such as cans, home furnishings or even cars. (Approximately 25% of a car body is recycled metal!) + **Building spoils** – such as soils or rubble are recycled to either fill quarries where a mineral was extracted or as hardcore for foundation support. + **Plastics** – often reused for further plastic products. + **Organic garden waste** – can be made into compost for gardeners to use. What is actually recycled is dependent on the local authority (council) responsible for waste. It doesn't take much to sort out packaging and waste into different materials for recycling. This stops much of our waste going into landfill sites.
Landfill	Landfill is simply waste that is buried in the ground. This is not good for the environment as it leads to ground and water contamination, where unseen dangerous products seep into the ground.
WEEE	Waste electrical and electronic equipment must, under the WEEE Regulations, be recycled as separate items from any other form of recycling. See page 12 in Topic 1.1 for more detail of WEEE Regulations.
Hazardous waste	Hazardous waste products must be treated with caution, processed and disposed of. Some contain hazardous gases or powders, which could be released if thrown into a skip and damaged when disposed of. As a result, many items – e.g. fluorescent tubes that contain mercury and phosphor powder – need to be disposed of in a special crusher that seals in the harmful substances. Always consult the manufacturers' documentation or seek advice from your local recycling centre.

Topic 5.2 Reporting of hazardous waste

 REVISED

Buildings and equipment contain hazardous waste and electricians commonly come across the following items listed in Table 1.28 below.

Table 1.28

Asbestos	See page 25 in Topic 3.7 for more detail.
Lead	Lead is harmful if heated (vapour) or filed/cut, forming dust particles. Lead can be found in the following products: + paint + batteries + solders. **What you should do**: + Wear PPE, respiratory aids and disposable coveralls to protect yourself when working with lead-based products. + Ensure you are completely free of any lead before eating etc.
Acids	Acids can burn skin as well as some containers when draining or transporting acids for disposal. Acids are commonly found in batteries, such as lead acid batteries. **What you should do**: + Acids should only be handled by persons trained to deal with them. + Always report the presence of acids to a supervisor or responsible person.
PCBs	Polychlorinated biphenyls (PCBs) were commonly used as an electrical insulator and can be found in older: + transformers + capacitors + fluorescent light fittings. Modern equipment containing PCBs is now labelled but older equipment was not. **What you should do**: + PCBs should only be handled by persons trained to deal with them. + Always report the presence of them to a supervisor or responsible person.

Mercury	Mercury was used extensively in electrical products and especially in electronic switches and LCD back-lit screens.
	What you should do:
	+ In small quantities such as these, products are disposed of under WEEE Regulations, where the mercury is safely removed.
	+ Higher quantities are found in mercury vapour lamps – care must be taken to keep these undamaged for recycling.

Topic 5.3 Type of pollution

REVISED ●

Strict regulations now restrict all forms of pollution. If pollution is noted, it should be reported to the relevant authority, which can then take the necessary actions to contain or stop it. Forms of pollution are listed in Table 1.29 below.

Table 1.29

Land contamination	Land can be contaminated by many site operations, such as fuel or oil spillage, or rubbish/hazardous waste being left lying and covered by soils etc.
	If land becomes contaminated, it must be reported immediately to the Environment Agency to take the necessary action.
Water contamination	Many rainwater drains run directly into water courses, such as rivers and streams. Spillages of fuels, oils, acids or bleaches can run into these water courses and cause terrible harm to wildlife and plants.
	If water becomes contaminated, it must be reported immediately to the Environment Agency to take the necessary action.
Air contamination	Air pollution is most commonly caused by fires, whether accidental or intentional. Many local councils have banned intentional fires due to the pollution caused and nuisance to others.
	If you see a bonfire, it should be reported to the local authority.
Noise	Noise is not always considered as pollution, but it can cause serious problems to people local to the source of the noise. This could be a continuous source of noise, such as machinery operating for long durations, or short but loud sources, such as radios playing loud music.
	Local authorities may have restrictions regarding noise levels at certain times of the day and night. They should always be consulted before construction projects begin.
Light	Light pollution could be large scale, as in the glow of lights in the night sky, to small-scale annoyance such as poorly directed outside lights on someone's home.
	Light pollution can be controlled by local council planning policies, as well as careful design of lighting systems and selection of luminaires.

> **Exam tip**
>
> Pollution doesn't have to be a physical contaminate – light and noise are also considered as pollution, especially when they cause a nuisance or obstruction.

LO6 The structure and roles of individuals and organisations within the construction industry

In any construction project, many different people are involved, carrying out many different roles. Some represent the client and their interests; some are simply there to build a building and make a profit. These roles and structures are easily described using diagrams showing the roles and relationships. They are illustrated by Figures 1.12, 1.13 and 1.14.

Topic 6.1 Types of site personnel

REVISED ○

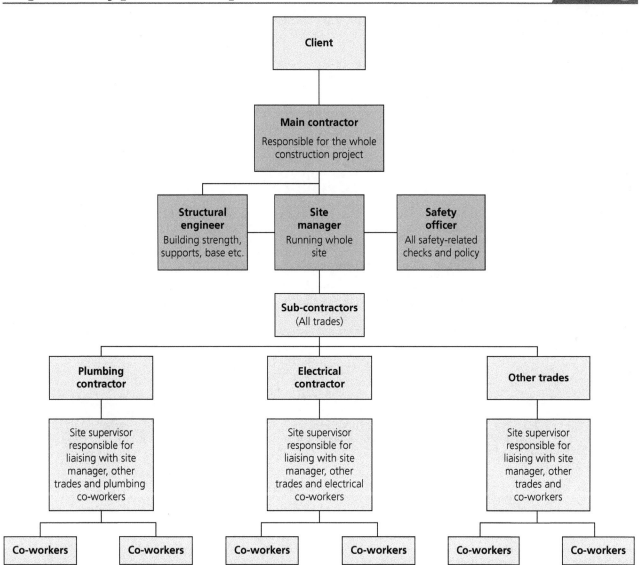

Figure 1.12 Roles and hierarchy of persons on a construction site

Topic 6.2 Client and representatives

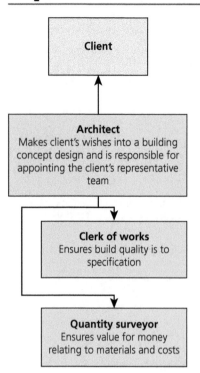

Figure 1.13 The client's representative team

Topic 6.3 Relationships in the contract structure

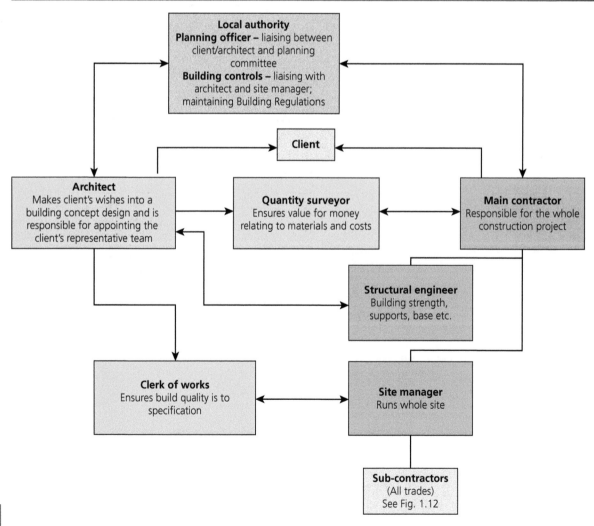

Figure 1.14 Relationship between all representatives

Check your understanding and progress at **www.hoddereducation.co.uk/myrevisionnotes**

Topic 6.4 Role of industry bodies

Within the electrical industry, there are many organisations, boards and schemes that help promote the industry, consumer rights and safety.

Institutes

The institute for the electrical industry is the Institute for Engineering Technology (IET), which is responsible for the following, amongst many other things:

+ Working in partnership with the British Standards Institute to form a joint committee known as JPEL 64. They are responsible for producing BS 7671 – the IET Wiring Regulations: Requirements for Electrical Installations.
+ Providing guidance publications giving a complete understanding of technical requirements.
+ Working with the Engineering Council to recognise qualifications and award levels of status. For example, on completion of a full electrical apprenticeship, you are able to apply for EngTech recognition.
+ Closely monitoring the industry, identifying skills gaps and opportunities.

Certificating organisations

CERTSURE, NAPIT, STROMA and NICEIC are all certificating bodies, forming part of the Competent Persons Scheme. The scheme providers monitor their member contractors to ensure high standards are met. They each maintain a register of electricians who are members and conduct regular inspections of their work, and update training and further continued professional development (CPD).

> **Continued professional development (CPD)** Electricians are expected to keep up to date with the latest technology advancements, as well as changes to regulations.

Electrical Safety First (ESF)

ESF is a charity with the mission to promote electrical safety for consumers. As well as publishing the latest safety information on issues such as product recalls for white goods, they also provide advice for consumers regarding electrical installation work, who should do the work and how to organise it.

> **White goods** Appliances used in the home, usually the kitchen, such as washing machines, fridges and dishwashers.

They also provide a series of best practice guides for electricians on how to achieve compliance with some aspects of regulations that cause the industry problems. To do this, they work closely with many industry partners such as certificating organisations, IET, City & Guilds and Electrical Contractors' Association (ECA).

Joint Industry Board (JIB)

The JIB is an organisation that works with employers to set, for their members, minimum pay grades, holiday entitlement and travel allowances.

As well as this, they set achievement grades based on knowledge and experience; for example, Electrician, Approved Electrician, Technician and Trainee Levels.

The JIB is also responsible for partnering the Electrotechnical Certification Scheme (ECS), which is the nationally recognised scheme for industry operatives by grade and occupation. They set the minimum training requirements for grades, such as Electrician. Their card scheme recognises those who progress onto further training by keeping these qualifications on their records and displayed on their members' cards. Their cards adopt a colour-code system, and most electricians aspire to gold card status.

Exam-style questions

1 Which of the following regulations are non-statutory?
- **a)** BS 7671: IET Wiring Regulations.
- **b)** Control of Noise at Work Regulations.
- **c)** Working at Height Regulations.
- **d)** Manual Handling Operations Regulations.

2 Which statutory regulations cover the handling of hazardous substances?
- **a)** NEBOSH
- **b)** COSHH
- **c)** EaWR
- **d)** PUWER

3 What **minimum** length of ladder is needed to safely access the roof area as shown in Figure 1?

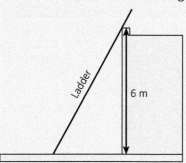

Ladder

6 m

Figure 1

- **a)** 1.50 m
- **b)** 6.10 m
- **c)** 7.18 m
- **d)** 9.32 m

4 How many of the top rungs of a ladder, or treads of a stepladder, is it recommended to avoid standing on when using?
- **a)** 1
- **b)** 2
- **c)** 3
- **d)** 4

5 What is an essential item of PPE when working on a high, narrow platform?
- **a)** Harness
- **b)** Pads
- **c)** Gloves
- **d)** Hi-vis vest

6 What is considered as the last line of defence when considering risk-reduction methods?
- **a)** Signage
- **b)** Evacuation
- **c)** Barriers
- **d)** PPE

7 What does a Class II power tool rely on as a means of electric shock protection?
- **a)** Earthing
- **b)** Bonding
- **c)** Insulation
- **d)** SELV

8 What colour indicates that a socket-outlet is 110 V?
- **a)** Violet
- **b)** Yellow
- **c)** Blue
- **d)** Red

9 What is a method statement?
- **a)** A set of regulations
- **b)** A set procedure
- **c)** A risk assessment
- **d)** A hazard assessment

10 The five steps to producing a risk assessment include:
- 1 Identify who may be harmed.
- 2 Identify the hazard.
- 3 Regularly review.
- 4 Evaluate the risk.
- 5 Record findings.

What is the correct order for carrying this out?
- **a)** 1 2 3 4 5
- **b)** 2 5 1 4 3
- **c)** 2 1 4 5 3
- **d)** 1 3 5 4 2

11 A worker is injured by twisting their ankle on a site and needs time off work to recover. After how many consecutive days off work is an organisation required to report the accident to the HSE?
- **a)** 2
- **b)** 5
- **c)** 7
- **d)** 10

12 What is an informal safety briefing on a construction site also known as?
- **a)** Toolbox talk
- **b)** Safety speech
- **c)** Drill discussion
- **d)** Site speaker

13 Which colour indicates a fire extinguisher that is safe to use on electrical fires?
- **a)** Red
- **b)** White
- **c)** Green
- **d)** Black

14 What type of signage is triangular, with a yellow background and black border and picture?
- **a)** Prohibition
- **b)** Warning
- **c)** Information
- **d)** Mandatory

15 Where in an old electrical system is asbestos **most** likely to be found?
- **a)** Cable insulation
- **b)** Rewireable fuse carriers
- **c)** Incandescent lamps
- **d)** Bathroom pull-cords

16 What HSE guidance publication should a voltage indicator used for safe isolation comply with?
  a) GS38
  b) L50
  c) INDG455
  d) GN8

17 What is the correct **minimum** number of tests needed to confirm the isolation of a three-phase circuit?
  a) 3 points of test
  b) 7 points of test
  c) 10 points of test
  d) 15 points of test

18 What product may contain lead?
  a) Paint
  b) Fuses
  c) Glue
  d) Resins

19 What is a quantity surveyor responsible for?
  a) The cost of a building project.
  b) The quality of a building project.
  c) The safety of all on-site operatives.
  d) The qualifications of all on-site operatives.

20 Which of the following organisations is an institute?
  a) NAPIT
  b) IET
  c) NICEIC
  d) JIB

2 Electrical science (Unit 202)

Scientific principles and an understanding of some maths are the backbone to good electrical work. Many regulations are based on scientific principles, such as Ohm's Law, and cable ratings are based on temperature effects of current and effects on conductor resistance.

This unit will remind you of the science and maths foundations needed to support your practical work and design techniques at this level.

LO1 Apply mathematical principles

Topic 1.1 Units of measurement; Topic 1.3 Work with geometry

REVISED

Electrical installation work involves a lot of calculations and formulae. Table 2.1 gives details of what is being calculated, SI units of measurement, and the formulae used to calculate the value in different circumstances. For calculating properties of shapes, refer to Figure 2.1 to see the values used.

> **SI units** The international system for units of measurement; these are the base values that should be used in any formula.

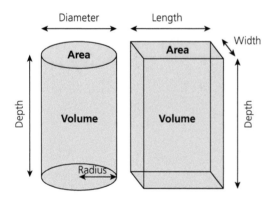

Figure 2.1 Calculating properties of shapes

Table 2.1

What needs calculating	SI unit of measurement	Symbol	How to work it out	Notes
Length	Metres (m)	L	The measurement from one point to another.	Length may also be called 'distance'.
Area (sometimes called 'cross-sectional area' or CSA)	Metres squared (m²)	A	$\text{length} \times \text{width}$	Applicable to a square or rectangle having four sides.
			$\dfrac{\text{length} \times \text{width}}{2}$	Applicable to right-angled triangles.
			$\pi \times \text{radius}^2 \ \text{or} \ \pi r^2$	Applicable to circles. The radius of a circle is half the value of the diameter.

What needs calculating	SI unit of measurement	Symbol	How to work it out	Notes
Volume	Metres cubed (m³)	V	$\text{length} \times \text{width} \times \text{depth}$ or $\text{area} \times \text{depth}$	Used for cubes or cuboids.
			$\pi r^2 \times \text{depth}$ or $\text{area} \times \text{depth}$	Used for cylinders.
Density	kg/m³	ρ	$\dfrac{\text{mass}}{\text{volume}}$	This depends on the material. For example, water has a density of 1000 kg/m³.
Weight	kg	F_g	$\text{mass} \times \text{gravity}$	Weight is the force acting on a surface due to gravity acting on the mass. Acceleration due to gravity is approximately 9.81 m/s² on Earth. (So weight and force are essentially the same thing!)
Force	Newtons (N)	F	$\text{mass} \times \text{acceleration}$	
Energy or work	Joules (J)	E	$\text{force} \times \text{distance}$ or $f \times d$	This is the method applicable to mechanical energy.
			$\text{power} \times \text{time}$ or $\text{watts} \times \text{seconds}$	This is the method used when calculating electrical energy.
Power	Watts (W)	P	$\dfrac{\text{energy}}{\text{time}}$ or $\dfrac{\text{joules}}{\text{seconds}}$	This is the method used to calculate mechanical power.
			$\text{volts} \times \text{current}$ or $V \times I$	This is the method used to calculate electrical power.
Velocity or speed	Metres per second (m/s)	v	$\dfrac{\text{distance}}{\text{time}}$ or $\dfrac{\text{metres}}{\text{seconds}}$	Velocity is the speed in a particular direction. Where the direction is not specific, such as up, down, north, etc., this is commonly known as speed.
Acceleration	Metres per second squared (m/s²)	a	$\dfrac{\text{velocity}}{\text{time}}$	Acceleration is the measurement of the rate of change of speed.

Worked example

A load, having a density of 1.3 kg/m³ and dimensions of 2 m length, 2 m width and a depth of 3 m, needs to be lifted a distance of 72 m in a time of 3 minutes. Calculate the following values:

a) Volume of the load
b) Mass of the load
c) Weight or force of the load
d) Energy needed to raise the load
e) Power needed to raise the load
f) Speed or velocity at which the load is raised

a) **Volume of the load**

$$\text{volume} = l \times w \times d$$

So:

$$2 \times 2 \times 3 = 12 \text{ m}^2$$

b) **Mass of the load**

$$mass = density \times volume$$

So:

$$1.3 \times 12 = 15.6 \text{ kg}$$

c) **Weight or force of the load**

$$force = mass \times acceleration$$

So:

$$15.6 \times 9.81 = 153 \text{ N}$$

> **Exam tip**
>
> Remember, acceleration here is due to the pull of gravity and is, on average, 9.81 m/s² on Earth.

d) **Energy to raise the load**

$$energy = force \times distance$$

So:

$$153 \times 72 = 11,016 \text{ J}$$

e) **Power to raise the load**

$$power = \frac{energy}{time}$$

So, as three minutes is equal to 3×60 seconds, this equals 180 seconds:

$$\frac{11,016}{180} = 61.2 \text{ W}$$

f) **Speed or velocity of the load**

As the load travels 72 m in 180 seconds upwards, the velocity is:

$$\frac{72}{180} = 0.4 \text{ m/s}$$

> **Exam tip**
>
> Some questions may use the lever principle to test your understanding of force, so remember:
>
> $$F \times d = F \times d$$

Now test yourself   TESTED ◯

1 A load, having a mass of 140 kg, is to be raised a distance of 80 m in 2 minutes. How much power is needed to complete this task?

We must use the SI unit of measurement within a formula, but because sometimes the values are very large or very small, we use indices.

> **Indices** A method of simplifying a large or small quantity by using a power of 10; for example:
>
> $$6 \times 10^{12} = 6,000,000,000,000$$

Table 2.2 below shows numbers expressed as indices (to the power of ten) including the prefix used to describe the values.

Table 2.2

Actual number	Number shown to the power of 10	Prefix used
1,000,000,000,000	10^{12}	tera (T)
1,000,000,000	10^{9}	giga (G)
1,000,000	10^{6}	mega (M)
1000	10^{3}	kilo (k)
100	10^{2}	hecto (h)
10	10^{1}	deka (da)
0.1	10^{-1}	deci (d)
0.01	10^{-2}	centi (c)
0.001	10^{-3}	milli (m)
0.000,001	10^{-6}	micro (µ)
0.000,000,001	10^{-9}	nano (n)
0.000,000,000,001	10^{-12}	pico (p)

To use these in a calculation, you can use the EXP or x10ˣ button on your calculator.

Check your understanding and progress at **www.hoddereducation.co.uk/myrevisionnotes**

Worked example

Using Ohm's Law, what is the power of a circuit having 45 MV and 50 kA?

So: 45 MV is 45,000,000 V (or 45×10^6 V)

And 50 kA is 50,000 A (or 50×10^3 A)

So, on a calculator: 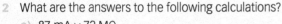 2.25 × 10¹² W, or 2.25 TW (or even 2,250,000,000,000 W).

Exam tip

If your calculator gives you an answer in a format that doesn't suit, use the ENG button, as this expresses the same answer into different indices. Use the button directly to move down in indices and shift ENG to move up the indices.

Typical mistake

Always double check your calculator when looking at results of calculations, as many students miss the ×10ˣ on the screen, which changes the answer significantly.

Now test yourself   TESTED ◯

2 What are the answers to the following calculations?
  a) 87 mA × 72 MΩ =
  b) $45 \times 10^{-6} \times 45 \times 10^6$ =
  c) 400 kV × 25 A =

Topic 1.2 Work with equations or formulae   REVISED ◯

One skill that is essential for success in exams, and when carrying out day-to-day calculations, is being able to transpose formulae. For the formulae we use in electrical installations work, there are some very simple steps to transposing a formula to calculate the quantity required.

> **Transpose** To rearrange a formula in order to make the value you need to find the subject of the formula.

Formulae involving adding and subtraction, including linear equations

+ The value to be found must be a '+' (positive).
+ Any value moved over the '=' sign becomes a '−' if it was a '+' or a '+' if it was a '−'.
+ Any value with **no** '+' or '−' in front is regarded as '+'.

Make 'c' the subject:

$$a + b - c = d$$

As c is a '−c' then it must be moved over the '=' sign to become a '+ c', so:

$$a + b = d + c$$

So, to leave 'c' as the only subject, move 'd'. Remember, as 'd' has no sign in front of it, it is thought of as '+ d', so:

$$a + b - d = c$$

We can prove this works by giving the values numbers and repeating the steps. For example:

$$30 + 40 - 10 = 60$$

$$30 + 40 = 60 + 10$$

$$30 + 40 - 60 = 10$$

> **Typical mistake**
>
> Multiple-choice questions that require a formula to be transposed will often have the three incorrect answers from using the wrong methods of transposing the formula. So, **never** accept an answer as correct because it is an answer option. Always double-check *your* transposition is correct!

39

Formulae involving multiplication and division

✛ Any value to be found must be on the top row.
✛ Any value moved over the '=' sign moves from bottom to top or top to bottom.
✛ Any value **not** showing as top or bottom is regarded as top.

Make 'g' the subject:

$$\frac{e \times f}{g} = h$$

As 'g' is at the bottom and must be at the top, it must be moved over the '=' sign, so:

$$e \times f = h \times g$$

So now 'g' is at the top, we need to move 'h' away to leave 'g' as the subject:

$$\frac{e \times f}{h} = g$$

Once again, using numbers:

$$\frac{100 \times 2}{50} = 4$$

$$100 \times 2 = 4 \times 50$$

$$\frac{100 \times 2}{4} = 50$$

Formulae involving square and square root

✛ To unlock values in a square root by removing it, you must square the other side.

Make 'b' the subject:

$$\sqrt{a^2 + b^2} = c$$

As 'b' is locked by the square root, we need to unlock it by squaring the other side of the '=' sign:

$$a^2 + b^2 = c^2$$

Using the rules of addition and subtraction, move the 'a^2' to leave the '$+b^2$' on its own. Remember, the 'a' has no '+' or '−' so it must be '+a', which becomes '−a' when moved:

$$+b^2 = c^2 - a^2$$

Now, we need to find 'b', **not** 'b^2'. To drop the square, we need to do the opposite on the other side, so:

$$b = \sqrt{c^2 - a^2}$$

Pythagoras theorem and trigonometry

Many electrical quantities use triangles and principles. Trigonometry and Pythagoras theorem are used to quantify the relationship between values and are principles you will see a lot, especially at Level 3.

Pythagoras theorem

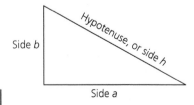

Side b

Hypotenuse, or side h

Side a

Figure 2.2 A right-angled triangle

Check your understanding and progress at **www.hoddereducation.co.uk/myrevisionnotes**

Pythagoras shows the relationship between the lengths of the three sides of a right-angled triangle:

$$h = \sqrt{a^2 + b^2}$$

If side 'a' was 2 m and side 'b' was 4 m, the hypotenuse can be found by:

$$\sqrt{2^2 + 4^2} = 4.47 \text{ m}$$

Trigonometry

Trigonometry uses angles within triangles to determine the length of sides, or determine angle values based on the lengths of the sides.

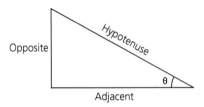

Figure 2.3 Right-angled triangle with angle marked θ (theta)

Depending on the value you need to find within the triangle, you can use any one of the following formulae depending on what values you have.

$$\sin\theta = \frac{\text{opposite}}{\text{hypotenuse}}$$

$$\cos\theta = \frac{\text{adjacent}}{\text{hypotenuse}}$$

$$\tan\theta = \frac{\text{opposite}}{\text{adjacent}}$$

Exam tip

To remember the three formulae, we can use the acronym 'SOH CAH TOA'.

SOH is $\text{Sin} = \dfrac{\text{Opp}}{\text{Hyp}}$;

CAH is $\text{Cos} = \dfrac{\text{Adj}}{\text{Hyp}}$;

TOA is $\text{Tan} = \dfrac{\text{Opp}}{\text{Adj}}$

Worked example

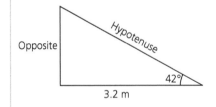

Figure 2.4

For the triangle shown in Figure 2.4, calculate the:

a) length of the side marked opposite

b) length of the hypotenuse.

a) To determine the length of the opposite side, we can choose the 'tan' formula as this involves the two known values, and the one we need to find:

$$\tan\theta = \frac{\text{opposite}}{\text{adjacent}}$$

And to transpose to make the opposite the subject:

$$\tan\theta \times \text{adjacent} = \text{opposite}$$

So:

$$\tan 42 \times 3.2 = 2.9 \text{ m}$$

b) To find the length of the hypotenuse, we can now use any of the other two formulae, or Pythagoras. In this instance, we will use Pythagoras.

$$h = \sqrt{3.2^2 + 2.9^2} = 4.3 \text{ m}$$

LO2 Understand direct current principles

Topic 2.1 Electron theory

REVISED ◯

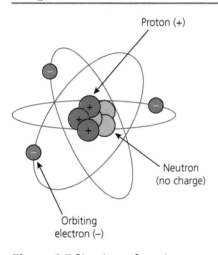

Figure 2.5 Structure of an atom

Figure 2.5 shows the structure of an atom, which includes:
+ the nucleus made up of protons, which are positively charged, and neutrons, which have no charge
+ electrons, which are negatively charged and orbit the nucleus.

Electricity is the flow of electrons from one atom to another. Materials that are good conductors have electrons which move out of orbit from atom to atom. When the material is connected to an electro-motive force (EMF) such as a battery, the flow can be controlled in one direction. This is because the electrons are attracted to the positive plate of the battery.

This flow of electrons is called charge and happens in materials that are good conductors, such as copper. Atoms which keep their electrons in orbit make good insulators, such as atoms in rubber or PVC.

What makes materials or elements different are the atoms. The number of electrons orbiting the nucleus is different for different types of materials. For example, copper has 29 electrons but iron has 26. The number of electrons in an atom equals the number of protons.

When 6.24×10^{18} electrons flow in one direction, this is one coulomb of charge.

Topic 2.2 Properties of an electrical circuit

REVISED ◯

Table 2.3 shows key electrical values, their SI unit of measurement, and the formulae used to calculate their value.

Table 2.3

What needs calculating	SI unit of measurement	Symbol	Associated formulae
Charge	Coulomb (C)	Q	$1\,C = 6.24 \times 10^{18}$ electrons
Current	Ampere (A)	I	$I = \dfrac{Q}{t}$ or $I = \dfrac{V}{R}$
Electro-motive force or circuit voltage	Volt (V)	V	$V = I \times R$
Resistivity	Ohm-metre (Ω-m)	ρ	See section on resistance below.
Resistance	Ohm (Ω)	R	$R = \dfrac{V}{I}$ or $\dfrac{\rho L}{A}$
Power	Watt (W)	P	$P = V \times I$ or $I^2 \times R$

Conductors and resistivity

Before we look at Ohm's Law in detail, we need to look at conductors and insulators, and how much resistance a circuit has based on the material used as the conductor. This is based on the resistivity (ρ) of the material.

As different materials have different numbers of electrons, they all conduct electricity differently. Each material has a resistivity value based on the measurement of the resistance of a 1 m³ block of the material at 20°C. Table 2.4 shows some common resistivity values.

> ### Check your understanding
>
> 2 What do the atomic numbers represent on a periodic table?

Current (*I*) The measure of charge over time (*t*) and measured in amperes (A). So:

$$I = \frac{Q}{t}\ (\text{A})$$

Resistance The measure of how well a cable or insulator conducts electricity in ohms (Ω). The lower the value of ohms, the better it conducts; the higher the ohms, the worse it conducts.

Ohm's Law The relationship of current, voltage and resistance in an electrical circuit:

$$V = I \times R$$

Table 2.4

Material	Resistivity value (ρ)	Material	Resistivity value (ρ)
Copper	0.0172×10^{-6} Ω-m	Hard rubber	1×10^{13} Ω-m
Aluminium	0.028×10^{-6} Ω-m	Glass (average value)	10×10^{12} Ω-m
Gold	0.024×10^{-6} Ω-m	Dry wood	1×10^{14} Ω-m
Steel (used in cables)	0.46×10^{-6} Ω-m	PVC (average)	1×10^{15} Ω-m

To work out the value of resistance, we use the resistivity calculation, which uses the material's length (*L*) and cross-sectional area (*A*), and apply:

$$R = \frac{\rho L}{A}$$

When calculating resistivity, if you always use:
+ a resistivity value in µΩ-m (micro-ohm-metres or ×10⁻⁶)
+ and the CSA in mm² (×10⁻⁶)

then the two values of ×10⁻⁶ cancel out. You can simply input the values without using the ×10⁻⁶.

> **Exam tip**
>
> Remember, when you see a formula with no sign between two values, multiply them. So ρL is actually $\rho \times L$.

> ### Worked example
>
> What is the resistance, at 20°C, for 30 m of copper cable having a cross-sectional area (CSA) of 2.5 mm²?
>
> $$R = \frac{\rho L}{A} \qquad \text{So: } R = \frac{0.0172 \times \cancel{10^{-6}} \times 30}{2.5 \times \cancel{10^{-6}}} = 0.21\ \Omega$$

Now test yourself   TESTED

4   Calculate the resistance of an aluminium cable at 20°C having a length of 100 m and a CSA of 25 mm².

Conductor resistance is calculated at 20°C – but resistance does change with temperature. If temperature is increased, so does resistance.

Table 2.5 shows how copper conductors change in resistance with a change in temperature.

Table 2.5

Temperature increase from 20°C	% change in conductor resistance	Factor used to change resistance at new temperature from 20°C	Temperature decrease from 20°C	% change in conductor resistance	Factor used to change resistance at new temperature from 20°C
+5°C	2%	1.02	−5°C	2%	0.98
+10°C	4%	1.04	−10°C	4%	0.96
+15°C	6%	1.06	−15°C	6%	0.94
+20°C	8%	1.08	−20°C	8%	0.92
+50°C	20%	1.20	−50°C	20%	0.80

Looking at the table, you can see that the resistance of copper changes by 2% for every 5°C change in temperature.

Check your understanding

3   Look at Table I2 in the IET On-site Guide and see how they have presented their temperature factors. Why are they going in the opposite way?

Worked example

What is the resistance of a copper conductor having a CSA of 1.5 mm² and a length of 46 m at a temperature of 70°C?

At a temperature of 20°C:

$$R = \frac{0.0172 \times 46}{1.5} = 0.53 \, \Omega$$

So, at 70°C:

$$0.53 \times 1.2 = 0.636 \, \Omega$$

Now test yourself   TESTED

5   What is the resistance of a 6 mm² copper conductor having a length of 80 m at a temperature of 15°C?

6   You have a cable reel that has been partly used and you are not sure how much cable is left on it. The cable is 1.5 mm² copper cable. When you measure the resistance of the cable, you get a reading of 0.55 Ω. How much cable is left on the reel?

When current passes through a conductor, heat is generated and this increases the resistance.

Check your understanding

4   Have a look at Table I1 in the IET On-site Guide and see how they have shown conductor resistances. Choose one or two cable sizes and check the values using the resistivity of copper. Are the values the same? If not, why do you think this is?

Insulators

Insulators are used to stop electric current flow as they are poor conductors. We use insulating materials to surround conductors to stop people getting electric shocks and to stop current leaking from one conductor to another.

In theory, insulators are materials that stop current. But that is only true if the force behind the current flow, voltage, is low enough. If you increase voltage, eventually the insulator will break down and current will flow.

The table shows common materials used for cable insulation, their maximum operating temperature and final temperature.

Maximum operating temperature The temperature above which the insulation resistance begins to break down, allowing current to flow with the circuit voltage.

Final temperature The point where the insulation starts to melt.

Table 2.6

Class of insulation	Material	Maximum operating temperature	Final temperature
Thermoplastic	Polyvinyl chloride (PVC)	70°C	160°C
	Polyethylene (PE)		
Thermosetting	Cross-linked polyethylene (XLPE)	90°C	250°C
	Ethylene propylene rubber		
Mineral	Magnesium oxide	*250°C	**1083°C

* This is normally limited to 105°C due to the risk of burns and fire rather than its insulation properties breaking down. But it can be installed in areas having high temperatures.

** This is the melting point of the copper conductors, not the point where the insulation breaks down.

We will look at insulation in more detail, such as why we choose certain insulators, in Chapter 3 of this book.

Sources of EMF and current flow

Electro-motive force (EMF) is the pressure that pushes the current around a circuit and is measured in volts.

There are three sources of EMF:
+ **Chemical:**
    + Using two chemicals in an electrolyte solution causes electrons to be drawn to one of the chemicals.
    + The reaction in the anode (chemical 1) releases electrons, and the reaction in the cathode (chemical 2) absorbs them.
    + The result is electricity when connected to a circuit, as the electrons flow from anode to cathode.
    + This will continue to produce electricity until one or both of the electrodes run out of the substance necessary for the reactions to occur.
    + This is called a cell – several cells placed together is called a battery.
+ **Magnetic:**
    + When a ferrous metal cuts through a magnetic field, an EMF is induced – it is the principle used by all rotating generators. (This will be covered further in Learning Outcome 3 on pages 51–59.)
+ **Thermal:**
    + When a junction of two materials has a difference in temperature on each side, this can produce an EMF – it is called a thermocouple.
    + This method is not as common as chemical and magnetic sources of EMF, as thermo-generators would need huge amounts of material to produce the amount of electricity made by a small magnetic generator (e.g. a wind turbine).

Ferrous metal A metal containing iron, which because of this can stick to a magnet. Metals that do not contain iron, such as aluminium, are known as 'non-ferrous' metals.

Check your understanding

5   Research online three chemicals used to make batteries or cells.

Voltage and current may be direct (DC) or alternating (AC). While the magnetic source of EMF can be either AC or DC, chemical and thermal sources are DC only.

When electricity was first being experimented with, it was thought that current flowed from the positive to the negative. This is called **conventional current flow**. In reality, electrons flow from negative to positive, which is called **electron current flow**.

Topic 2.3 Principles of an electrical circuit

To work out the quantities of a DC circuit, we apply Ohm's Law:

$$V = IR$$

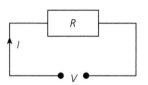

Figure 2.7 Simple DC circuit

Figure 2.7 shows a simple circuit with a resistor (R), a current (I) and an EMF (V).

We can apply Ohm's Law to calculate values based on those we know.

> **Exam tip**
>
>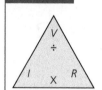
>
> **Figure 2.6**
> To remember Ohm's Law, use the triangle in Figure 2.6 and cover the value you wish to find. It will then leave you the two other values and tell you if they need to be divided or multiplied.

Worked example

If the circuit in Figure 2.7 had a voltage of 12 V and a resistance of 8 Ω, what would be the current?

$$V = IR$$

So, transposed to find the current:

$$I = \frac{V}{R}$$

So:

$$\frac{12}{8} = 1.5 \text{ A}$$

Now test yourself

7 If the circuit in Figure 2.7 had a current of 6 A, and a voltage of 120 V, what would the circuit resistance be?

Series circuits

When resistances in a circuit are connected one after the other, they are connected in series.

The total resistance is found by adding all the resistances together:

$$R_{total} = R_1 + R_2 + R_3$$

and so on.

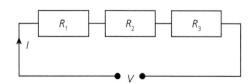

Figure 2.8 Circuit having three resistors in series

The circuit shown in Figure 2.8 has a supply voltage of 200 V and three resistors of 10 Ω, 25 Ω and 15 Ω in series. Calculate the circuit current.

As the current is based on the total circuit resistance, this needs to be determined first:

$$R_t = 10 + 25 + 15 = 50 \ \Omega$$

Now Ohm's Law can be applied to determine the circuit current:

$$I = \frac{200}{50} = 4 \text{ A}$$

TESTED ○

8 What is the circuit current if a circuit had a supply voltage of 400 V and 12 Ω, 8 Ω, 16 Ω and 22 Ω resistors in series?

In a series circuit, the current remains the same through each resistor but the voltage across each resistor is different. The voltage across each resistor can be calculated by applying Ohm's Law to that resistor. The value of all the voltages must equal the total circuit voltage.

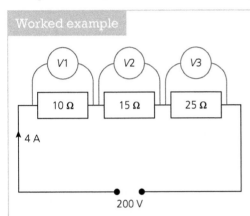

Figure 2.9 Measuring voltage in a series circuit

To calculate the voltage across each resistance:

$$V1 = 4 \times 10 = 40 \text{ V}$$

$$V2 = 4 \times 15 = 60 \text{ V}$$

$$V3 = 4 \times 25 = 100 \text{ V}$$

And to check the values equal the circuit voltage:

$$40 + 60 + 100 = 200 \text{ V}$$

TESTED ○

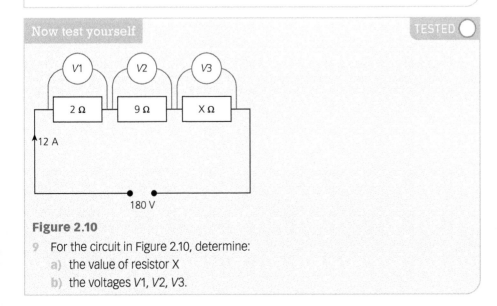

Figure 2.10

9 For the circuit in Figure 2.10, determine:
a) the value of resistor X
b) the voltages V1, V2, V3.

Parallel circuits

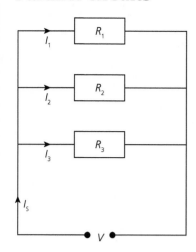

Figure 2.11 Parallel resistors in a circuit

When resistors are arranged in parallel (see Figure 2.11), the supply voltage remains constant across all resistors – but the current in each branch changes.

The value of the current in each branch is based on the values of resistance and voltage in each branch. The total of each branch will equal the supply current (I_s).

The total resistance for the circuit is calculated by:

$$\frac{1}{R_{total}} = \frac{1}{R_1} + \frac{1}{R_2} + \frac{1}{R_3} \quad \text{and so on.}$$

A golden rule is that the total resistance must be lower than the lowest resistor in the circuit.

Worked example

Calculate the total resistance and each value of current, if the circuit in Figure 2.11 has the following values:

+ $R_1 = 20\,\Omega$
+ $R_2 = 30\,\Omega$
+ $R_3 = 40\,\Omega$
+ $V = 400\,V$

$$\frac{1}{R_{total}} = \frac{1}{20} + \frac{1}{30} + \frac{1}{40} = \frac{1}{0.108\ldots} = 9.23\,\Omega$$

$$I_1 = \frac{V}{R_1} = \frac{400}{20} = 20\,A$$

$$I_2 = \frac{V}{R_2} = \frac{400}{30} = 13.33\ldots\,A$$

$$I_3 = \frac{V}{R_3} = \frac{400}{40} = 10\,A$$

$$I_s = \frac{V}{R_{total}} = \frac{400}{9.23} = 43.33\ldots\,A$$

And to check:

$$20\,A + 13.33\ldots\,A + 10\,A = 43.33\ldots A$$

Typical mistake

When calculating parallel resistances, many students forget the total is divided into 1 at the end. Using the example, they might think the result is 0.108 Ω. Make sure you do the final calculation to get the correct result.

Check your understanding and progress at **www.hoddereducation.co.uk/myrevisionnotes**

TESTED ○

10 The circuit in Figure 2.11 has the following values:
+ $R_1 = 5\,\Omega$
+ $R_2 = 100\,\Omega$
+ $R_3 = 200\,\Omega$
+ $V = 100\,V$

Calculate the total circuit resistance and each current value.

11 A 10 m run of steel conduit contains a 1.5 mm² copper CPC which is connected to the earth of a socket-outlet at the end of the conduit run. The conduit has a 20 mm outside diameter and an 18 mm internal diameter (the bit the cables go in).

As the conduit acts as an earth in parallel to the CPC, what is the overall resistance of the conductor?

To work out the total resistance on a calculator, use the x^{-1} button. So, for the values in the worked example, push:

Power in parallel and series circuits

Before we look at calculating the power in series and parallel circuits, let's look at a circuit with series and parallel resistors, then calculate the power.

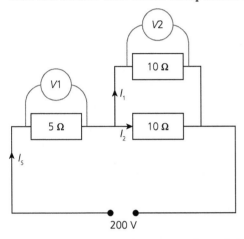

Figure 2.12 Circuit with series and parallel resistances

Using the circuit in Figure 2.12, the current in the 5 Ω resistor is the same as the supply current (I_s). The current through each 10 Ω resistor is calculated in the same way as a parallel circuit and must total the supply current. The voltage V2 is the same across each 10 Ω resistor and V1 + V2 must equal the supply voltage.

Using the circuit in Figure 2.12, calculate:
+ Total resistance of the parallel section (R_p)
+ Total circuit resistance
+ Circuit current I_s
+ Voltage $V1$ and $V2$
+ Currents I_1 and I_2

$$\frac{1}{R_p} = \frac{1}{10} + \frac{1}{10} = \frac{1}{0.2} = 5\,\Omega$$

As the parallel branch totalling 5 Ω is in series with the 5 Ω resistance, the total resistance of the circuit is:

$$R_{total} = 5 + 5 = 10\,\Omega$$

So I_s can be found using Ohm's Law:

$$I_s = \frac{V}{R_{total}} = \frac{200}{10} = 20\,A$$

Voltage $V1$ is then found by:

$$V1 = I_s \times R = 20 \times 5 = 100\,V$$

Voltage V2 will also be 100 V as the total resistance of the parallel section is also 5 Ω.

Current I_1 is found by applying Ohm's Law to that section:

$$I_1 = \frac{V2}{10} = \frac{100}{10} = 10\,A$$

Current I_2 has the same properties:

$$I_2 = \frac{V2}{10} = \frac{100}{10} = 10\,A$$

As the total circuit current is 20 A, the current is split evenly over the two equal resistances in parallel.

49

Power dissipated in a circuit can be calculated in two ways.

Either:

$$P = V \times I$$

Or:

$$P = I^2 R$$

Worked example

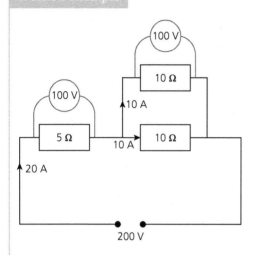

Figure 2.13 Calculating power in a circuit

Figure 2.13 shows the circuit from the previous example with the calculated values. Remember that the total circuit resistance was 10 Ω.

To work out the power of the whole circuit, we can use the total circuit or supply values, so:

$$P = V \times I = 200 \times 20 = 4000 \text{ W or } 4 \text{ kW}$$

Equally:

$$P = I^2 R = 20^2 \times 10 = 4000 \text{ W or } 4 \text{ kW}$$

So, any formula can be used, depending on the values known.

The power dissipated in the 5 Ω resistor is:

$$20 \text{ A} \times 100 \text{ V} = 2000 \text{ W or } 2 \text{ kW}$$

The power dissipated in one of the 10 Ω resistors is, using the other formula:

$$10 \text{A}^2 \times 10 \text{ Ω} = 1000 \text{ W}$$

As the other 10 Ω resistor is the same, a total of 2000 W is dissipated in the parallel part of the circuit.

Now test yourself   TESTED ◯

12 Using the circuit shown in Figure 2.14, calculate:

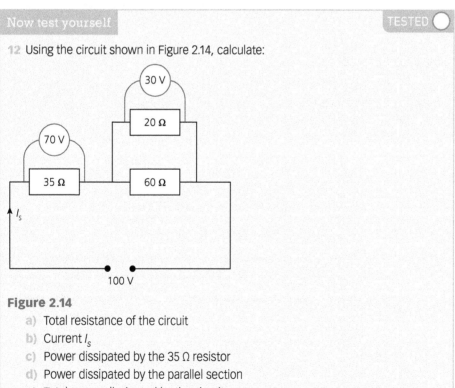

Figure 2.14

- a) Total resistance of the circuit
- b) Current I_s
- c) Power dissipated by the 35 Ω resistor
- d) Power dissipated by the parallel section
- e) Total power dissipated by the circuit

Topic 2.4 Measurement of electrical circuits

The instruments listed in Table 2.7 below are used to measure electrical quantities and are connected as shown in Figure 2.15.

Table 2.7

Quantity to be measured	Instrument used	Connection	Notes
Voltage	Voltmeter	In parallel to the item being measured	The instrument is measuring the voltage difference from one side of the load to the other. This is called potential difference.
Current	Ammeter	In series with the load	Ammeters can measure small currents this way but for much larger currents, a current transformer is needed (see Topic 3.4).
Resistance	Ohmmeter	In parallel to what is being measured	This will only work if the circuit or item being measured is disconnected from any power source.
Power	Wattmeter	Both parallel and series	This measures the voltage and current. It calculates the resulting power.

> **Potential difference** The difference in voltage from one terminal to another.

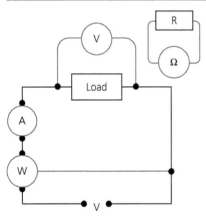

Figure 2.15 How instruments are connected to measure circuit quantities

> **Induced** To force the creation of (in electrical terms), i.e. a current will force the creation of a magnetic field, and so a magnetic field will force the creation of a current.
>
> **Electro-magnetism** Where a magnetic force is produced by passing current through a conductor. When a direct current is passed through a coiled conductor, this produces the same effect as a bar magnet with magnetic poles.

LO3 Understand electromagnetic properties

Topic 3.1 Principles of magnetism

Magnetism is key to electricity. Current in a conductor produces magnetic fields. Magnetic fields can also be used to induce current.

To understand magnetism, we need to look at the SI units linked to permanent magnets and electro-magnetism, as well as the formulae used to calculate quantities.

Table 2.8

What needs calculating	SI unit of measurement	Symbol	How to work it out
Magnetic flux	Weber (Wb)	ϕ	Amount of magnetic flux in a magnet or flux change around a current-carrying conductor.
Induced electro-motive force (EMF) (generator principle)	Volts (V)	E	$$E = \frac{\phi}{t} \text{ or}$$ $$E = \beta Lv$$ t = time in seconds v = velocity a conductor is rotated in a magnetic field L = length of conductor in a magnetic field
Magnetic flux density	Tesla (T)	β	$$\beta = \frac{\phi}{A}$$ A = area in m² containing the flux
Force produced by a field reacting with a current-carrying conductor (motor principle)	Newtons (N)	F	$$F = \beta LI$$ L = the length of a conductor in the field I = current in the conductor
Electromagnetic induction	Henry (H)	L	Quantity given to a coil (e.g. a transformer coil) for its ability to induce (produce) a voltage: + back into itself (self-induction) + into another coil (mutual induction).

Permanent magnets

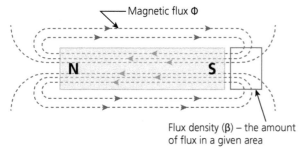

Magnetic flux Φ

Flux density (β) – the amount of flux in a given area

Figure 2.16 Permanent bar magnet showing lines of flux and flux density

+ A permanent magnet is a ferrous metal that has become magnetised and produces a rotating magnetic field of flux.
+ Permanent magnets have poles called North and South. The flux travels from North to South outside the magnet.
+ When two magnets are placed together, they either attract or repel, depending on the position of the poles.
+ Magnets attract when opposite poles are placed close to one another, as the flux in each magnet rotates in the same direction. They pull together, as shown in Figure 2.17.

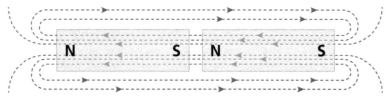

Figure 2.17 Two bar magnets attracting as the lines of flux rotate together, becoming one magnet

Check your understanding and progress at **www.hoddereducation.co.uk/myrevisionnotes**

When two poles that are the same are placed together, the force of the flux pushes the magnets apart, as the rotation is against each other. The flux becomes squashed in a small area. If the magnets are not fixed, the force will cause movement.

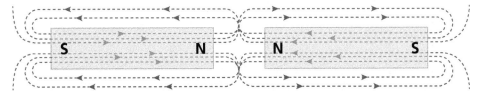

Figure 2.18 Two bar magnets repelling as the lines of flux are opposing and become squashed where the magnets meet

Electro-magnetism

+ When a current passes through a current-carrying conductor, a magnetic field is induced around that conductor. The flux rotates in a direction relative to the current. Figure 2.19 shows the direction of flux around a conductor cross-section.

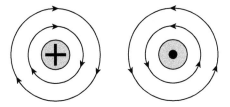

Figure 2.19 Lines of flux rotating around a current-carrying conductor

+ The cross indicates the current is moving away and the dot indicates current moving towards. Think of it like a dart moving through the air, with flights at the back (moving away) and a point at the front (moving towards). The amount of flux induced is relative to the amount of current.

Exam tip

To remember the direction of flux lines, use the screw rule. If you were tightening up a screw using a cross-head screwdriver, you would turn it clockwise.

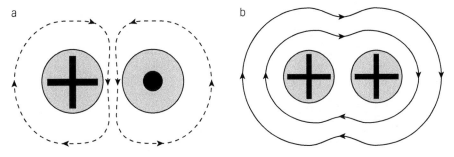

Figure 2.20 Cancellation (a) and totalling (b) effects of current-carrying conductors

+ When two conductors are placed together having equal current, but current in each is flowing in the opposite direction, the flux induced by each will cancel each other out and the field will not form.
+ This is common in twin cables having a line and neutral, as the cancellation means a reduction in magnetic interference.
+ If the flux lines move in the same direction, they will join and increase in intensity. This is used to an advantage when creating an electro-magnet or solenoid.
+ Coiling a conductor around a tube and passing a current through the conductor would create an electro-magnet.
+ This is because the current is moving in the same direction at any point in the coil.

Solenoid A conductor wound into a tight helix to produce an intensified magnetic field.

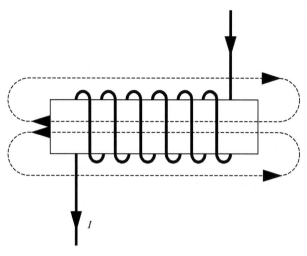

Figure 2.21 A solenoid or electro-magnet showing the direction of flux paths

+ The strength of the magnetic field can be increased by:
  + putting in more turns
  + increasing the current.

> **Exam tip**
>
> To remember the polarity of a solenoid, use the grip rule. Using your right hand, use your fingers to indicate current direction and your thumb will point to the North pole.
>
>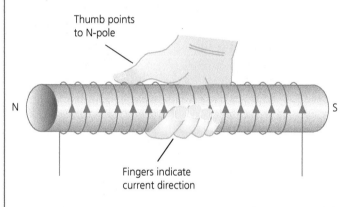
>
> **Figure 2.22** The grip rule

Solenoids are used widely in electrical equipment, especially for switching large currents using smaller ones. The coil of a solenoid does not need too much power to establish a strong enough magnetic field to force the movement of a switch contact that switches a larger powered circuit. Devices that do this are called contactors and relays.

Topic 3.2 Conductors in magnetic fields REVISED

+ If a current-carrying conductor is placed into a permanent magnetic field, the field induced by the conductor will react with the permanent field.
+ Like repulsion, force will move the conductor.
+ The amount of force, in newtons, is found by:

$$F = \beta L I$$

where:
  + β = the flux density of the permanent field in tesla
  + L = the length of conductor in the field in metres
  + I = the current flowing in the conductor in amperes.

Calculate the force on a 10 cm conductor in a magnetic field of flux density 0.5 T, when 5 A flows through the conductor.

Before calculating, remember 10 cm is 0.1 m:

$$F = \beta LI \text{ so } 0.5 \times 0.1 \times 5 = 0.25 \text{ N}$$

This is the principle of how motors work, but instead of using permanent magnets to create the magnetic field, electromagnets are commonly used.

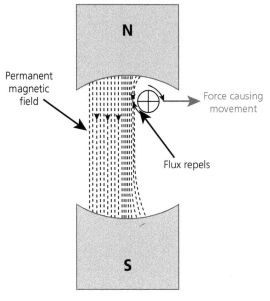

Permanent magnetic field

Force causing movement

Flux repels

N

S

Figure 2.23 Current-carrying conductor in a magnetic field

The direction of the force can be determined using Fleming's 'left-hand rule' for motors. The direction of current flow in a generator can be determined using Fleming's 'right-hand rule' (remember: GeneRIGHTers).

✛ If a conductor with no current was rotated at a given velocity within a magnetic field, this would induce an EMF onto the conductor.
✛ The amount of EMF can be calculated by:

$E = \beta Lv$ where:

✛ E = EMF in volts
✛ L = length of conductor in the main magnetic field in metres
✛ v = velocity of the conductor moving in the field in metres per second.

Calculate the EMF induced into a 3 m conductor rotated at 4 m/s in a magnetic field having a flux density of 0.75 tesla.

$$E = \beta Lv \text{ so } 0.75 \times 3 \times 4 = 9 \text{ V}$$

TESTED

13 A 1.5 m coil of cable is within a magnetic field having a flux density of 1.2 tesla.
   a) Calculate the force if the conductor had 6 A passing through it.
   b) Calculate the EMF if the conductor was rotated at 6 m/s.
14 If a generator had a diameter of 0.5 m and the rotor rotated at 30 revolutions per second, what is the rotor velocity?

Topic 3.3 Principles of electrical generation

REVISED

If a conductor was forced to rotate inside a magnetic field, such as the one shown in Figure 2.24, the EMF and current induced would flow in one direction when the conductor is at the top, and another direction when at the bottom. This is how alternating current (AC) is induced.

55

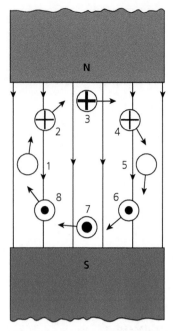
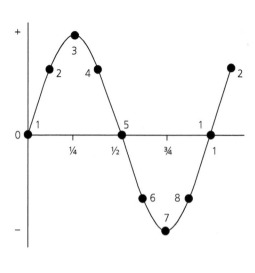

Figure 2.24 The EMF generated per rotation. The time taken to complete one rotation is the periodic time of the waveform shown.

The waveform produced by generating AC is called a sine wave. It has several properties which are shown on Figure 2.25.

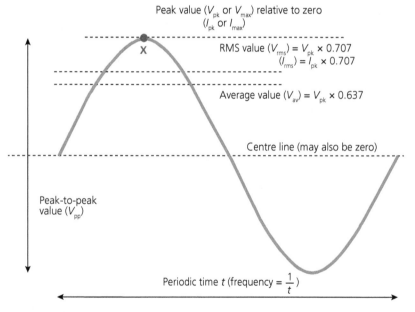

Figure 2.25 The different properties of a sine wave

Topic 3.4 Transformer principles

Transformers are used to change or transform current and voltage in an AC circuit.

There are several different types of transformer. Table 2.9 shows what they are and their typical uses.

Table 2.9

Transformer	Typical applications
Isolation	♦ Electric shock protection by providing electrical separation. ♦ The windings are completely separate. ♦ Current is induced on the secondary side by **mutual induction**. ♦ These transformers may be step-up (voltage increases on the output side) or step-down (voltage decreases on the output side).
Auto	♦ Power transformers as the secondary side is tapped off the primary, so fewer copper windings. ♦ Works using the principle of **self-induction**. ♦ These transformers are step-down only.
Current	♦ Measuring current in conductors. ♦ The primary winding is the single conductor being measured. ♦ The current in the secondary winding is induced by mutual induction of the magnetic field, which surrounds a conductor carrying current. ♦ The more current in the conductor, the greater the field strength. ♦ These transformers are step-up only to reduce the current.

Mutual induction Where one winding produces a magnetic field and that magnetic field cuts through a second winding, producing a current in that second winding.

Self-induction Where one winding produces a magnetic field and that magnetic field induces a current back into the same winding.

Induced To force the creation of (in electrical terms), i.e. a current will force the creation of a magnetic field, and so a magnetic field will force the creation of a current.

Basic operating principles of transformers

When current is passed through a conductor, a magnetic field is induced. This field is channelled by the core, so the flux cuts through the secondary winding, which induces current.

The voltage and current output is dictated by the ratio of primary to secondary turns.

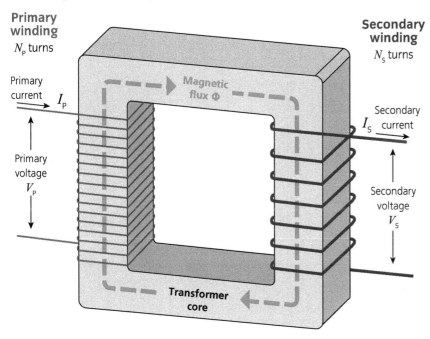

Primary winding
N_P turns

Primary current I_P

Primary voltage V_P

Magnetic flux Φ

Secondary winding
N_S turns

Secondary current I_S

Secondary voltage V_S

Transformer core

Figure 2.26 Simple transformer arrangement

Types of transformer core

There are two common types of core:
♦ Shell: more efficient as the flux is concentrated in both directions.
♦ Core: flux loss is higher, but cheaper to manufacture.

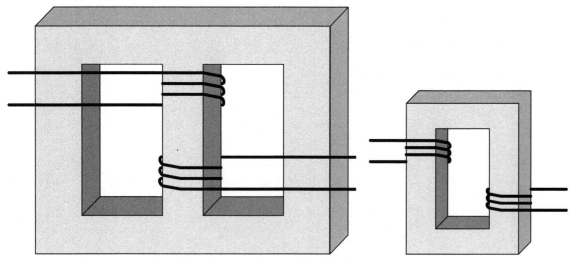

Figure 2.27 Shell-type core (left) and a core-type core (right)

The purpose of the core is to channel the magnetic flux induced by the windings to where it is needed.

Cores are laminated, meaning they are made from very thin, insulated slices of metal. This creates a high electrical resistance, which reduces eddy current (see below) flow which acts as a loss. As magnetic flux is not impeded by insulation, the flux path remains effective.

Transformer losses

There are two main types of loss in a transformer.

Iron loss

Iron loss can occur in two ways:

+ **Eddy current loss:**
    + Induced current in the core which rotates around the core, creating heat.
    + By laminating the core, eddy current flow is restricted and reduces.
+ **Hysteresis loss:**
    + Due to the material used for the core.
    + More power is needed if a material requires more energy to magnetise it, then de-magnetise (due to alternating current changing direction).

Copper loss

This is due to the resistance of the copper windings. The greater the resistance, in relation to the current flow, the greater the power loss:

$$P = I^2 R$$

> **Laminated** Where a block or object is made up of many layers. In the case of a transformer, these layers are bolted together to form the core.
>
> **Eddy current** A current that is induced into a metal by the magnetic field. These currents rotate around the metal (like eddy currents in a stream causing mini whirlpools) and these currents cause the material to heat up. As using energy to produce heat is a loss, reducing eddy currents reduces this energy loss.

> **Worked example**
>
> If the winding of a transformer has a resistance of 0.4 Ω and the current passing through the winding is 15 A, determine the copper loss.
>
> $$P = 15^2 \times 0.4 = 90 \text{ W}$$

> **Now test yourself**   TESTED ◯
>
> 15 A transformer has a copper loss of 120 W when 30 A flows through it. What is the resistance of the winding to produce this loss?

Turns, voltage and current relationships

Use this formula to determine the relationship between voltage, current and the number of turns within a transformer:

Check your understanding and progress at **www.hoddereducation.co.uk/myrevisionnotes**

$$\frac{N_p}{N_s} = \frac{V_p}{V_s} = \frac{I_s}{I_p}$$

Transformers are often given ratios for the number of turns primary (N_p) to the number of turns secondary (N_s).

So, if a transformer had one turn on the primary side for ten on the secondary, the transformer would be in the ratio of 1:10.

Typical mistake

Notice in the transformer ratio formula that the current values are inverted. Many forget this in the exam!

Worked example

A transformer has 400 turns on the primary side and 80 on the secondary. If the input voltage was 230 V and the primary current was 3 A, determine the:

a) secondary voltage

b) secondary current

c) transformer ratio.

$$\frac{400}{80} = \frac{230}{V_s} = \frac{I_s}{3}$$

a) To determine the secondary voltage, we do not need the current part of the formula, just the number of turns and voltage.

$$V_s = \frac{230 \times 80}{400} = 46 \text{ V}$$

b) To determine the secondary current, we do not need the voltage part of the formula.

Exam tip

Use the given values each time. Although the voltage values determined in a) could now be used to determine the current values in b), if you made a mistake in a), b) would also be wrong.

$$I_s = \frac{400 \times 3}{80} = 15 \text{ A}$$

c) To determine the ratio, we simply divide the turns values, so:

$$\frac{400}{80} = 5$$

So as the primary has more turns, the ratio would be 5:1.

Now test yourself

TESTED ◯

16 Complete the table with the missing transformer characteristics using the formula:

$$\frac{N_p}{N_s} = \frac{V_p}{V_s} = \frac{I_s}{I_p}$$

	Voltage	Current	Turns
Primary	11 kV		300
Secondary	250 V	100 A	

Exam tip

You must be able to transpose formulae for all examinations.

LO4 Understand electronic components

Topic 4.1 Operating principles of components; Topic 4.2 Applications and uses of components

REVISED ◯

In the electrical industry, experience working with electronic devices becomes less frequent. The ability to identify and replace faulty components becomes far less routine, but a full circuit board replacement is more common.

However, you do need to be able to identify the component symbols and have a brief understanding of what they do and where you may find them.

Table 2.10 gives a brief outline of what each component does and Figure 2.28 shows the different symbols.

Table 2.10

Component name	Function	Examples of application/usage
Resistor	Used to add resistance to a circuit to limit current or voltage.	+ One of the most common components. + Found in all electronic circuits.
Capacitor	Used to store a charge and release when required.	+ Used to smooth out rectified circuits (see Figure 2.29 for rectification) and for power factor correction.
Thermistor	A resistor where resistance changes with temperature.	+ Used in sensor or indicator circuits where temperature change requires monitoring (e.g. monitoring water temperature in a boiler).
Light-dependent resistor (LDR)	A resistor where resistance changes with light.	+ Photoelectric cells where light levels are monitored and lights switch on when dark.
Diode	Allows current to flow in one direction only.	+ Commonly used in rectifiers changing AC to DC (see below). + Smaller power diodes are referred to as signal diodes.
Zener diode	Allows current to flow in one direction only when a particular voltage is reached.	+ Used to monitor system voltage in electronic circuits. + Often referred to as a voltage regulator.
Photo diode	Acts like a diode allowing current flow in one direction when light is detected.	+ Used commonly in control circuits or sensors (e.g. pick-up sensors for remote control devices or data sent by fibre optics).
Light-emitting diode (LED)	A diode that gives out light when current passes through in one direction.	+ Very common devices used for indicator lights and for all types of general lighting.
Transistor	Used to amplify current. When a small current is detected by the base terminal, it allows a larger current to flow from collector terminal to emitter terminal.	+ Used very commonly to amplify a signal. + Where a small signal, such as a microphone signal, is inputted onto the collector of the transistor, the transistor will amplify the signal into a much higher one for the loudspeakers.
Diac	This is two zener diodes arranged to allow an alternating current flow, once a set voltage is reached.	+ Diacs are commonly used with a triac in dimmer circuits or motor control circuits such as soft start motor controls.
Triac	This is a three-lead device which acts as a voltage-driven switch. It switches on and off rapidly with frequency. This can be used to chop a waveform acting as a voltage limiter.	+ The diac will regulate voltage to the dimmer circuit and the triac gate is fed by a variable resistor. + The triac is then used to chop a voltage waveform feeding the light or motor, relative to the voltage pulsed onto the gate.

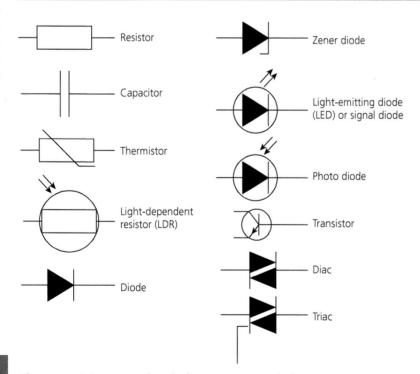

Figure 2.28 Common electrical component symbols

Check your understanding and progress at **www.hoddereducation.co.uk/myrevisionnotes**

One type of circuit where diodes are used is to rectify AC to DC.

The circuit shown in Figure 2.29 uses four diodes to change the AC into a DC by flipping the sine wave to only positive cycles.

The capacitor, which stores charge when the voltage peaks, releases the charge when the voltage falls to smooth the voltage to pure DC.

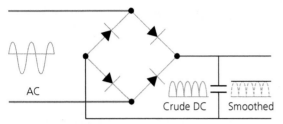

Figure 2.29 Full wave rectification of AC to DC

Exam checklist

+ SI units
+ Using and transposing formulae
+ Using indices (to the power of)
+ Trigonometry and Pythagoras
+ Calculating areas and volumes
+ Atomic structure
+ Resistivity and conductors
+ Temperature effects on conductors
+ Properties of insulators
+ Electrical quantities and Ohm's Law
+ Series circuits
+ Parallel circuits

+ Measuring circuit values
+ Magnetism, repulsion and attraction
+ Electromagnetism and solenoids
+ Force and EMF calculations
+ How an AC sine wave is induced
+ Sine wave properties
+ Transformer cores
+ Transformer ratios
+ Transformer losses
+ Transformer types and uses
+ Electronic component principles and symbols

Exam-style questions

1 What is the area of the circle shown in Figure 1?

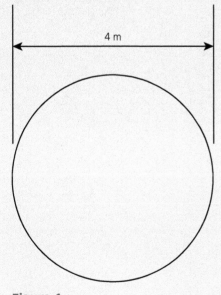

4 m

Figure 1

a) 3.144 m^2 c) 16.00 m^2

b) 12.56 m^2 d) 50.26 m^2

2 Which is the SI unit of measurement for density?

a) kg/m   c) kg/m^3

b) kg/m^2 d) kg/m^{10}

3 How much energy is needed to raise a 150 kg mass 9 m?

a) 132 J

b) 1350 J

c) 13,243.5 J

d) 13,500.5 J

4 Which value is the same as 48×10^{-6} ohms?

a) 48 MΩ c) 48 mΩ

b) 48 kΩ d) 48 $\mu\Omega$

5 What is the correct transposition of the formula shown in Figure 2 to make L the subject?

$$R = \frac{\rho L}{A}$$

Figure 2

a) $L = \dfrac{\rho R}{A}$ c) $L = \dfrac{RA}{\rho}$

b) $L = \rho RA$ d) $L = \rho + R + A$

6 Which is the unit of measurement for charge?
a) Farad   c) Ampere
b) Ohm   d) Coulomb

7 What is the length of a 4 mm² copper conductor having a resistance of 0.28 Ω at 20°C? Resistivity of copper is 0.0172 μΩ-m.
a) 1.24 m   c) 65.1 m
b) 45.6 m   d) 88.3 m

8 How does temperature affect a conductor?
a) An increase in temperature will increase conductor resistance.
b) A decrease in temperature will increase conductor resistance.
c) A decrease in temperature will have no effect on conductor resistance.
d) An increase in temperature will have no effect on conductor resistance.

9 What is the maximum operating temperature of a thermoplastic PVC-insulated cable?
a) 30°C   c) 70°C
b) 50°C   d) 110°C

10 Three resistors are connected in series and each has a resistance of 20 Ω. The circuit has a supply voltage of 180 V.
What is the total circuit current?
a) 3 A   c) 60 A
b) 6 A   d) 3600 A

11 Two resistors of 5 Ω and 20 Ω are connected in parallel.
What is the combined resistance?
a) 2.5 Ω   c) 0.4 Ω
b) 4.0 Ω   d) 25 Ω

12 Which statement is true regarding a circuit containing two resistors in series?
a) The voltage across each resistor is the same as the supply voltage.
b) The voltage across each resistor is three times the supply voltage.
c) The current through each resistor is the same as the supply current.
d) The current through each resistor is three times the supply current.

13 A circuit has a resistance of 50 Ω and a current of 2 A. How much power is dissipated?
a) 25 W   c) 100 W
b) 50 W   d) 200 W

14 What happens to the magnetic flux around a current-carrying conductor when a second conductor of equal current, but flowing in the opposite direction, is placed next to it?
a) The magnetic field is cancelled out.
b) The magnetic field strength doubles.
c) The magnetic field strength halves.
d) The magnetic field strength remains unchanged.

15 What is the unit of measurement for magnetic flux density?
a) Weber   c) Tesla
b) Laminate   d) Henry

16 How much force acts on a conductor carrying 25 A in a field of 0.2 T where 0.2 m of conductor is in the field?
a) 1 N   c) 100 N
b) 10 N   d) 1000 N

17 Which number, in Figure 3, shows the correct location of the North pole of the electro-magnet?

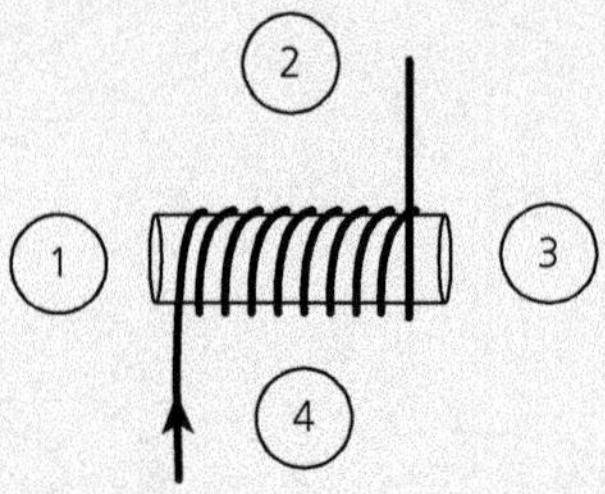

Figure 3
a) 1   c) 3
b) 2   d) 4

18 What type of loss is reduced in a transformer core by laminating the core?
a) Henry flow losses
b) Copper losses
c) Eddy current losses
d) Hysteresis losses

19 A transformer has a turns ratio of 20:5. What is the output voltage if the input voltage is 200 V?
a) 5 V   c) 500 V
b) 50 V   d) 5000 V

20 Which electronic component acts as an amplifier?
a) Resistor   c) Transistor
b) Diode   d) Capacitor

Check your understanding and progress at **www.hoddereducation.co.uk/myrevisionnotes**

3 Electrical installation (Unit 203)

This chapter looks at the practical part of being an electrician. We will take another look at:
+ tools commonly used by electricians;
+ how to erect cable containment/management systems;
+ wiring and terminations.

You will apply a lot of this in your practical synoptic assessment but your exam will still test your understanding of the different aspects of practical work. You will need to identify components, fixings and types of termination, as well as understand procedures for installing equipment.

LO1 Use tools commonly used in electrical installation practices

There is only one topic in this learning outcome but it covers a wide range of tools and equipment.

Topic 1.1 Use tools for electrical installation

REVISED

In this topic, we will list the common tasks and tools needed to carry out that task. The topic is broken into three areas:
+ common tasks and hand tools used
+ measurement and marking tools
+ power tools.

Table 3.1 shows the most common tasks, but an electrician may need to carry out many more.

Table 3.1

Task	Material/product	Tools used	Further information
Form	Conduit – metal	Vice and bender	+ Specific tool used for bending and threading conduit. + The former is changed to suit the size of conduit.
	Conduit – PVC	Bending spring	+ Sized for the conduit being bent. + The spring slots into the conduit to stop the walls collapsing.
	Tray	Tray bender	+ Specific tool used to bend metal tray to form internal and external bends.
Fabricate	+ Tray + Trunking	+ Vice and blocks + Hacksaw	+ Tray and trunking is fabricated by cutting and slotting using a hacksaw while held in a vice. + A block is used inside the trunking to stop the trunking from deforming in the vice.

> **Former** The shaped part of the bender that the conduit rests in when it is bent to keep its shape.

63

Task	Material/product	Tools used	Further information
Fabricate (continued)	Conduit	+ Vice and bender + Stocks and dies	+ Metal conduit requires threading, which is done using a die held in a stock. + The die cuts the thread into the metal.
Cut	Metal PVC or plastic	Hacksaw	+ Hacksaws are universal, but remember the teeth per inch (TPI). + The more teeth, the finer the cut and this is best for softer materials.
	Cable	+ Cable cutters + Side cutters	+ Side cutters are the general tool used to cut cables/cable cores up to 10 mm². + After that, more heavy-duty cable cutters are required.
	Wood	+ Cross-cut saw + Floorboard saw	+ A cross-cut saw is a general purpose saw for wood and some other materials. + A floorboard saw is specifically shaped to cut a single floorboard while it is still in place to create an opening in the floorboards.
	Plasterboard	Keyhole saw	+ Used to cut out small but accurate holes in plasterboard for recessed ceiling lights or accessories in plasterboard walls.
Fix	Nails or tacks	+ Claw hammer + Cross-pein pin hammer	+ The claw end of this hammer is used to lever out nails. + Cross-pein hammers have smaller heads and a thin wedged end, which is used to start nails off while holding them in your fingers.
	Ties	Cable tie gun	+ This tool will tension and cut the cable tie without crushing cables and leaving sharp cuts.
Strip	Sheathing	+ Knife and side cutters + Rotary flex stripper	+ Used to gently cut the outer sheath without cutting the inner insulation.
	Insulation	Cable strippers	+ There are several types but generally they are intended to strip back insulation without scoring the copper conductor.
	Armouring	Hacksaw	+ Used to score the steel armouring so the strands break, leaving an even, clean edge.
	MICC	MICC stripper and ringing tool	+ There are several types of stripper. They peel back the copper sheath to the mark/score left by the ringing tool.

Check your understanding

1 You have been asked to purchase some hacksaw blades for use on a site where a range of materials need cutting with a hacksaw. What are the different common TPI available for hacksaw blades?

Exam tip

Aluminium is a soft metal, so you might assume that a larger TPI should be used: but, as aluminium shavings get stuck in fine blades, a lower TPI is more suitable (e.g., 10–14 TPI).

MICC A mineral-insulated copper-clad cable. The mineral used for the cable insulation performs well at very high temperatures, making this cable suitable for high temperature applications (e.g. fire alarm systems where performance needs to continue in the event of a fire).

Check your understanding and progress at **www.hoddereducation.co.uk/myrevisionnotes**

Task	Material/product	Tools used	Further information
Terminate	Screw	Screwdriver	+ Screwdrivers come with several different types of head such as flat, cross-head or torx (see Figure 3.1). ![torx, hex, cross/pozi, slotted] **Figure 3.1** Screw heads from left to right: torx, hex, cross/pozi, slotted + Screwdrivers used by electricians will usually be insulated or VDE.
	Compression	Crimp tool	+ Used to joint conductors or terminate conductor ends with lugs. The crimp tool compresses or squashes the metal lug onto the conductor.
	Solder	Soldering iron	+ The hot soldering iron melts solder, which binds the terminals together.
	Displacement	Insulation displacement (ID) tool	+ This termination tool pushes the insulated conductor onto bladed terminals, which cut through the insulation, making contact with the conductor.
	MICC	+ Pot seal crimp + Pot wrench	+ The pot wrench threads the pot onto the cable copper sheath. Once the pot is filled with compound, the seal is fixed on using the seal crimp.
Tighten	Nuts/bolts	+ Spanners + Sockets	+ Spanners or sockets should always be used to ensure the nut isn't damaged. + Sockets use a ratchet tool. + Extension bars enable use in tight spaces where a spanner will not turn.
	Bushes	Bush spanner	+ A special type of spanner which grips a bush in a tight accessory. + Used when a traditional spanner or socket will not fit – it grips internally as well as externally.
	Screws	+ Screwdrivers + Bit heads	+ Screwdrivers come with several different types of head such as flat, cross-head or torx (see Figure 3.1). + Bit heads allow interchangeable heads where a bit holder screwdriver or powered screwdriver is used.
	Lock-rings	Grips or pipe grips	+ Grips (or pipe grips) can adjust to suit the size of ring.

VDE A European testing institute that gives insulated tools their safety certification to rigorous standards of safety. Certified VDE tools carry the VDE mark (see Figure 3.2).

Figure 3.2 The VDE standard mark

Now test yourself

1 You have been asked to obtain solder wire from a wholesaler and they ask you if you want wire with or without flux. What is flux and why does solder wire contain flux?

2 You need to purchase some 2-core 1.5 mm² MICC cable for a small lighting circuit. The client likes the look of the bare copper-type for effect in their restaurant where the lighting circuit is to be installed. MICC cable is categorised in a particular way. What should you ask for at the wholesaler so they know exactly what you want?

TESTED

Task	Material/product	Tools used	Further information
Grip/hold	To tighten	+ Grips or pipe grips + Spanners + Wrench/stillsons	+ Pipe grips are ideal for holding circular items such as pipe. + Wrenches (or stillsons) can grip or turn pipe fittings.
	To form/fabricate	+ Vices + Formers	+ Vices are used when cutting material to form bends, such as in trunking. + Formers are used to shape items, such as bending conduit.
Punch/mark	Metal	Centre punch	+ A centre punch, when hammered, will indent the material – e.g. it will mark and indent metal, giving a mark or starting point to drill without slipping.
File	Sharp edges	+ Files + Reamer	+ There are many different shaped files and their grades and patterns affect how fine or coarse the file cut is. + Reamers are used to file internal sharp edges and burrs in conduits. + Twisting the reamer files the tube inside of the cut section.

Exam tip

Some questions ask what tool is **most** suitable for a particular task. All options may be used but only **one** is most suitable. Make sure you read all options in the exam, not just the first option!

Measuring and marking tools

These are listed in Table 3.2 below.

Table 3.2

Tool	What it does
Spirit level	+ Ensures items are level horizontally or vertically when the bubble is centred in the tube (e.g. accessories or boxes with a flat surface).
Water level	+ Ensures two items apart from each other are at the same level. + For example, two water filled tubes are connected by a hose. As water will always keep to the same level, the same height can be found even in different rooms with different floor levels (see Figure 3.3). **Figure 3.3** A water level is used to find the same height in different locations
Laser level	+ Used to show a level line around a room by spinning a laser light 360° which projects the image of a line onto the walls all around the room. + Ensures trunking or conduit remains the correct level and height when installing around a room or along a wall.
Chalk line	+ A string is kept in a chalk-filled container. When pulled out, the chalk remains on the string. + When the string is held tight between two points and then pulled and released or twanged, the string leaves a chalk line on the wall. + This gives a straight-line mark for clips or saddles to follow.

Check your understanding and progress at **www.hoddereducation.co.uk/myrevisionnotes**

Tool	What it does
Rule	✦ Used to measure short but fine measurements on a surface. ✦ Scaled rules are useful for measuring scaled drawings – they convert the scale of the drawing dimensions to the actual distance for you.
Tape measure	✦ A long rule rolled into a compact container. ✦ When rolled out, it is used to measure longer distances. ✦ Steel tapes – 3 m to 5 m in length; nylon tapes – up to 25 m.
Measuring wheel	✦ Used to measure very long distances. ✦ The circumference of the wheel is usually 1 m. ✦ The revolutions of the wheel measure the distance. ✦ They are only suitable for firm ground, such as roads or short grass.
Range finder	✦ Optical devices with two lenses that use trigonometry to work out long distances to a marker. ✦ When viewing the marker through the lens, the finder is adjusted so the second lens angle is moved into focus where the two images overlap perfectly. ✦ Using the angle of the second lens, and the distance between the lenses, the distance to the marker can be calculated by tables (or the device itself).
Square	✦ Used to mark a straight edge at exactly 90° to another edge. ✦ Means items, such as trunking, can be cut perfectly square.

Power tools

These are listed below in Table 3.3.

Table 3.3

Tool	Uses
Drills	There are many types of power drill with different functions, depending on what needs drilling: ✦ Most portable drills will have drill and hammer drill functions. ✦ Some have a hammer only when chiselling, rather than rotary drilling. ✦ It is important to use the correct drill for the right application.
Drill bits	There are many types of drill bits that are used for drilling different materials and hole sizes, including the following: ✦ HSS: sharp-ended bits for drilling metals or plastics. ✦ Masonry: have a flatter head to the drill which is used to impact against **masonry**, breaking it as it drills through. ✦ Hole saws: circular tubes with cutting teeth at the top end used to cut bigger holes through metal or plastics, such as 20 mm diameter, but can be as large as 100 mm. ✦ Core bits: used for cutting large diameter holes in masonry walls. ✦ Augers: wide drill bits with a screw-like tip for cutting holes in wood. Usually for cutting larger or deeper holes than wood bits. ✦ Wood or spade bits: flat blades with a sharp tip used to centre the hole. They are used for cutting holes up to 25 mm diameter through wood.
Power saws	There are several types of power saw or cutter, including the following: ✦ Circular saws: for cutting long, straight lines through wood or sheet materials. ✦ Jigsaws: for cutting detailed lines which change in direction, such as circles. They can be used to cut out sections in wood or sheet metal by drilling a small hole first, then inserting the jigsaw blade to cut out a shape within the sheet material; e.g., if a long slot needs cutting into the side of a trunking to allow many cables to enter the trunking. ✦ Angle grinders: used to cut masonry or metals. They use a wheel but the blade is toothless and wears quickly. It will not grab like teeth can, making them suitable for harder materials.
Chase cutters	These are used for cutting **chases** into masonry walls to lay in cables and accessory boxes where **flush accessories** are required.

Masonry Material such as brick cement, concrete or plaster.

Chase A channel cut into a masonry wall to a particular depth to lay in cables or boxes so they will be buried in the wall and not seen.

Flush accessories Items, such as switches or socket-outlets, where all that is seen on the surface is the face plate. The rest is buried in the wall.

LO2 Erect cable containment/ management systems used in electrical installation

This Learning Outcome is mainly aimed at your practical skills when installing electrical components and equipment. In this section, we will look at the planning and selection aspects that may feature in your exam.

Topic 2.1 Selection of systems used in installation work; Topic 2.4 Install systems

REVISED

In these topics, we look at why particular systems are used. Topic 2.4 requires you to install a range of systems – but before they are installed, we need to understand why they are selected in the first place. As Topic 2.4 requires them to be installed, that is something you will practise at your centre and these skills will be assessed as part of your synoptic assessment.

There are several types of cable containment system or cable management system and these are selected based on their intended capacity and environment. These include those listed in Table 3.4.

Cable containment systems Systems that completely surround the cable(s), such as trunking.

Cable management systems Systems that support the cable(s), such as tray.

Table 3.4

System	Type	Uses
Conduit and ducting	There are several types and finishes of conduit: + rigid PVC + rigid metal + flexible	+ **PVC** is lightweight and does not corrode like metal. It can be white for internal use or black for internal and external. White is not suitable outside as it degrades in sunlight. + **Metal** provides good mechanical impact protection. It can be galvanised for corrosion protection or black enamelled for internal use. + **Flexible** conduit is either PVC or metal ribbed, giving good flexibility for connections to machines where vibration may be an issue. In all cases, cables need to be drawn through conduit. If cables are added later, there is a risk of damaging existing cables.
Trunking	Like conduit, trunking can be PVC or metal, depending on the level of protection required. Its shape also affects selection, as it may be used to house accessories, such as socket-outlets or switches.	Types of trunking include: **1 Dado**: + often in a D-shape + installed horizontally in the middle of a wall in offices to house power outlets, data outlets and telephone points + normally has two or three compartments inside it. **2 Standard cable**: + square or rectangular in shape + used to carry cables only with one internal compartment + it is very strong and fixed together with screws, giving a rigid installation.

System	Type	Uses
Trunking (continued)	It has several internal compartments for the segregation of different systems, such as data, power or communication cables.	**3 Lighting**: + metal box-shaped trunking which clips together + usually suspended in lines with light fittings mounted onto it + useful where uniform lighting is required, such as in a warehouse. **4 Mini**: + PVC trunking that is small + box-shaped + usually used for one or two cables only and gives a neat finish on a wall. **5 Skirting**: + like dado, skirting trunking is shaped to look like a skirting board + has a shaped top and flat bottom to look unobtrusive. **6 Floor**: + floor trunking is wide but shallow-boxed trunking is used for installing into a concrete screeded floor + this allows for access boxes containing sockets and accessories in an open floor area + usually made of metal and may have several compartments + cables can be added to the trunking at any time, as they can be laid in once the lid is removed (except for floor trunking). See Figure 3.4. Dado trunking Standard boxed cable trunking Skirting trunking **Figure 3.4** Standard trunking profiles
Tray	Tray is perforated to enable cables to be tied to it. Allows good air circulation, which manages heat in the cable.	+ Tray can support many cables but cables must have mechanical protection, such as a sheath, as this system only supports cables and doesn't contain them. + Useful for installing cables at different times as there is no lid and cables can be laid on rather than drawn in. See Figure 3.5. Perforated cable tray section Profiles showing vertical installation (A) and horizontal installation (B) **Figure 3.5** Perforated cable tray section and profiles

> **Segregation** A technical term used in BS 7671 meaning to keep different systems apart from each other to reduce the risk of interference or danger where different voltages are used.

My Revision Notes: City & Guilds Level 2 Advanced Technical Diploma in Electrical Installation (8202-20)

System	Type	Uses
Basket	Basket is a metal mesh used to support multiple cables. Cables are dropped into it rather than dressed and tied (like tray).	+ Basket can support many cables. + Cables must have mechanical protection such as a sheath as this system only supports cables, and doesn't contain them. + Useful for installing cables at different times – there is no lid and cables can be laid on rather than drawn in.
Ladder	Usually metallic galvanised and is a much stronger version of basket.	+ Cable ladder is usually used to support large cables, such as steel-wire-armoured (SWA) cable. + It is a sturdy system that supports much more weight than tray or basket without buckling or bending.

Check your understanding

3   What are Band I and Band II circuits and why are they segregated?

The different types of system are selected, depending on the following factors:

+ capacity
+ suitability (as described in Table 3.4)
+ external influences (e.g. environment, utilisation, building).

Capacity

This applies to conduit and trunking as they are containment systems – regulations require them to have an amount of free space within them to stop cables getting too hot. Other systems are not enclosed in this way, so their capacity is more affected by weight or physical size.

The maximum capacities of trunking and conduit can be worked out using the tables in Appendix E of the IET On-site Guide (OSG), which you are allowed to take into your exam.

Conduit capacity

Conduit capacities are worked out based on two situations:

1   If the conduit is a short straight run.

Or:

2   If the conduit has bends between inspection boxes.

Table E1 of the IET On-site Guide (OSG) gives factors for cables, depending on their size and type such as solid or stranded. Table E2 of the OSG gives the capacities for short straight runs of conduit, depending on the conduit diameter.

> **Exam tip**
>
> You can put sticky page-finder tabs in your On-site Guide to find tables and other key information quickly. City & Guilds exam rules allow this.

> **Inspection boxes** Conduit boxes where access can be made to cables and cables can be drawn in at these points.

Worked example

What is a suitable size conduit to house the following stranded cables?

+ $2 \times 2.5 \text{ mm}^2$
+ $3 \times 1.5 \text{ mm}^2$
+ $2 \times 4 \text{ mm}^2$

Looking at Table E1 in the OSG, the following factors are used for the different stranded cable sizes:

+ 1.5 mm^2: 31
+ 2.5 mm^2: 43
+ 4 mm^2: 58

The total for all these cables is:

$$(2 \times 43) + (3 \times 31) + (2 \times 58) = 295$$

Looking at Table E2 in the OSG at the conduit capacities, a 16 mm diameter conduit has a factor of 290.

As this is less than 295, this is **not** suitable, as the conduit would be too cramped. A 20 mm conduit has a capacity of 460 so this would have sufficient capacity.

So, a 20 mm conduit is selected.

Check your understanding and progress at **www.hoddereducation.co.uk/myrevisionnotes**

Now test yourself

TESTED ⬤

3   How many 2.5 mm² stranded conductors will fit into a 25 mm short straight conduit?

Exam tip

Make sure you use the bracket function on your calculator as the answer could be completely different if you do not!

Where a conduit is considered to be a long straight run (over 3 m) or the run has bends in it between inspection boxes, Tables E3 and E4 in the OSG are used. Table E3 gives the same factor for solid and stranded cables.

Worked example

What size conduit would be suitable for the following cables, where the conduit run is 5 m and incorporates two bends?
+ 3 × 1.5 mm²
+ 9 × 2.5 mm²
+ 3 × 4 mm²

Looking at Table E3 in the OSG, the following factors are used for the different stranded cable sizes:
+ 1.5 mm²: 22
+ 2.5 mm²: 30
+ 4 mm²: 43

The total for all these cables is:

$$(3 \times 22) + (9 \times 30) + (3 \times 43) = 465$$

Looking at Table E4 in the OSG at the conduit capacities, cross-referencing a 5 m run having two bends, a 25 mm conduit has a factor of 358, which is too small.

So, a 32 mm conduit must be selected as this has a factor of 643, which is larger than 465.

Check your understanding

4   You need to screw a bent piece of conduit into a threaded conduit box that is fixed to the wall as it forms part of a further conduit system. As the piece of conduit which needs to be screwed in has a 90° bend in it, you cannot turn it. How do you join the two items together if neither can be moved?

Now test yourself

TESTED ⬤

4   A conduit having a length of 3 m will have two bends and is expected to take the following cables:
+ 9 × 1.5 mm²
+ 6 × 2.5 mm²

What conduit size is required?

Typical mistake

Table E4 has lots of information in it so it is very easy to cross-reference it incorrectly – this is a common mistake. Sometimes, using a ruler or piece of paper to keep a line in focus can help.

The wrong options in a multiple-choice question will probably be the most common wrongly referenced answers, so always double check. Just because the answer you see is an option, it doesn't make it right!

Trunking capacity

Where trunking is used, Tables E5 and E6 are used the same way as conduit – but the difference with trunking is that bends are not taken into account at all. This is because cables are not pulled through trunking as they are in conduit.

However, what is different is the factors for cable don't only change if they are solid or stranded, but also by the type of insulation such as thermosetting and PVC.

Worked example

How many single-phase lighting circuits can be wired into a 50 mm × 50 mm trunking, assuming the lighting circuits were wired using 1.5 mm² stranded conductors having PVC insulation, and there were three conductors per circuit?

Referencing Table E6 and a 50 × 50 mm trunking, the factor is 1037.

Referencing Table E5, 1.5 mm² conductors have a factor of 8.6.

$$\frac{1037}{8.6} = 120.6 \text{ so } \frac{120.6}{3 \text{ cables}} = 40.2 \text{ or } 40 \text{ circuits}$$

So, the 50 × 50 mm trunking can house 40 different lighting circuits, each wired using 3 × 1.5 mm² conductors for their line, neutral and earth.

5 Where a 20 mm conduit has a 90° bend formed into it, what is the minimum permitted inner radius of that bend?

External influences

There are three general categories of external influence given in BS 7671 – these have an impact on selecting containment or management systems.

These are:

+ environment
+ utilisation
+ building.

Environment

This category covers many influences. The most common ones that impact on selection of systems are listed in Table 3.5.

Table 3.5

Environmental factor	Further information
Presence of water	+ In some installations, water may be present (e.g. rain, sprays from showers, waves in marinas). The protection the system offers against these actions is important. + For example, where water jets are present, PVC conduit may protect against water getting in – the conduit is watertight and any inspection boxes can use rubber gaskets with a lid. + If metal conduit is used in the presence of water, such as rain, the conduit must have a galvanised finish.
Impact	+ Impact may refer to industrial locations or storage areas where items are being moved frequently. + Containment and management systems need to be robust when in a location where damage may be caused by impact. + Metal is much more robust than PVC.

Ambient temperature	+ This is the temperature of the air surrounding a component.
	+ When it varies, systems will expand and contract:
	  + PVC expands at a greater rate than steel
	  + steel expands at the same rate as concrete.
	+ It is useful when concrete uses steel reinforcement, so wiring systems buried in concrete would also be best if they were steel.
	+ When PVC conduit is used in high and varying ambient temperatures, expansion couplers should be used – or the conduit will expand and buckle.

Corrosion	+ Water can cause corrosion to metal – so any metallic system used in outdoor locations needs to be galvanised to protect it. + Other corrosive substances include acids (e.g. acids in animal urine in farms) which can corrode metals. + Also, corrosive dusts such as lime in cement can cause metals to corrode – PVC may be a suitable alternative.

Vibration	+ Where equipment or locations move or vibrate, systems need to be flexible so they are not shaken and damaged.
	+ Flexible conduits are good in these situations, such as connections to motors or machines.
	+ Other corrosive materials that can damage systems are:
	  + metal-to-metal electrolytic corrosion where metal systems are fitted to metal surfaces
	  + plaster, cement and lime in untreated walls can cause metal to deteriorate
	  + oak or similar timber which can seep acids, causing metal corrosion.

Solar	+ Solar radiation from direct sunlight can degrade white PVC and make it very brittle. + When PVC conduit is installed outdoors, it should be coloured black to stop this happening.

5   When equipment is installed in an area where water is present, it needs suitable
    IP ratings to stop the ingress of the water (you will cover this in more detail at
    Level 3). What does IPX4 mean?

Utilisation

Think about who is using the building or what is happening in the building.
As an example, if there is a particular risk, such as fire caused by flammable
dust particles found in woodworking factories, the system must be dust-tight.

Building

The building itself is a consideration when choosing systems. Some things to
consider include:
+ What is the building constructed of (e.g. wood, which is combustible)?
+ Will the structure move (e.g. a jetty for mooring boats moves with waves
  and tide and so wiring needs flexibility)?
+ Is there the potential for the spread of fire in a building through ducts,
  such as large trunking? In these situations, fire barriers in the trunking
  must be installed.

6   BS 7671 uses a code system for all external influences. The code is made of
    two letters and then a number, such as 'AD5'. What do the letters and numbers
    represent?

Topic 2.2 Forming and fabricating containment systems

REVISED

If all containment systems were straight, installation would be easy.
Unfortunately, wiring must reach many parts of a building – so containment
systems must bend to fit the building.

Figure 3.6 shows the different types of bends, sets and junction/inspection
boxes found in conduit systems.

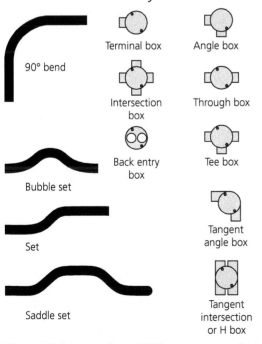

Figure 3.6 Types of conduit bends, sets and inspection boxes

Trunking has the same types of sets, with the exception of a bubble set, but selecting the right type of bend is important. Figure 3.7 shows the different types of trunking bends and junctions.

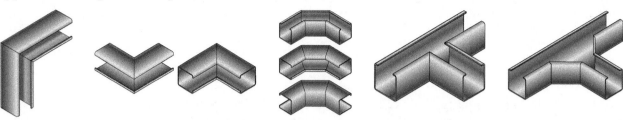

Figure 3.7 Types of trunking bends or junctions; from left to right: flat bend, external and internal bends, internal and external relieved bends, tee piece, relieved tee

Topic 2.3 Selecting fixings

Things to consider when selecting the correct fixing for a cable containment or management system include the following:

+ **Fixing systems**: the correct type of fixing for the system, such as saddles for conduit, brackets and threaded rods etc. for trunking and tray.
+ **Load bearing**: can the structure support the weight of the system?
+ **Aesthetics**: it must look right for its surroundings.
+ **Protection**: check if it is liable to damage by impact or other means, such as fire.
+ **Building fabric**: what methods are needed to secure to the building materials? For example, rawl plugs/screws, girder clamps for steel joists or woodscrews for timber.

When fixing conduit, there are four main options of saddle which secure the conduit to a surface (see Figure 3.8).

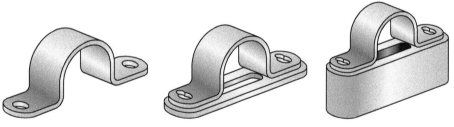

Figure 3.8 Types of saddle; from left to right: strap, spacer bar, distance and hospital

When installing conduit saddles, the spacing of the saddle is important to ensure it is well supported and looks pleasing.

Table D3 in the IET On-site Guide gives maximum distances between supports – depending on whether the conduit is rigid, metal, rigid PVC or pliable. Other factors that affect spacing include if the conduit is running up a wall (vertical) or along a wall (horizontal), and the overall diameter of the conduit.

> **Pliable** Means flexible or bendable. In this case it means flexible.

> **Worked example**
>
> A vertical 20 mm metal conduit needs to be fitted to a wall. What is the maximum distance between saddles?
>
> Looking at Table D3 in the On-site Guide, a 20 mm conduit falls into the $16 < d \leq 25$ category, as the conduit diameter (d) is greater than 16 mm but less than 25 mm.
>
> Cross-referencing this with rigid metal vertical (column 3), the maximum spacing is **2.0 m**.

Trunking can be supported by drilling and fixing straight through the back of the trunking onto a surface. To protect the cables inside, round-headed screws should be used – countersunk screws may give a sharp edge.

Trunking and tray can be supported by brackets, which space them off a wall, or by suspending them from steel rods on hanging brackets.

The maximum distance between supports for trunking can be found in Table D4 of the OSG. This is based on the overall trunking cross-section area, type of trunking, and whether the trunking is vertical or horizontal.

Typical mistake

Sometimes when Table D3 is referenced, people do not see the less than or equal to sign ($\leq$). They may assume a 25 mm diameter conduit falls into the next category in column 1 of the table. Make sure you study the table and understand the symbols used.

LO3 Install wiring systems and supports used in electrical installation activities

There is often confusion between a wiring system and a management or containment system.

Sometimes a containment system might be referred to as a 'wiring system'. However, technically the wiring system is the type of cable used and how it is fitted to a management system.

In this Learning Outcome, it is easier to look at each type of wiring system, why it is used or selected, and how it is supported, including maximum support distances.

Topic 3.1 Factors affecting the selection of wiring systems

REVISED

When selecting a wiring system, there are common factors that impact on the decision to use them, including the following:
+ **Cost** to install, including labour costs and time. Some systems are labour intensive and therefore expensive to install. For example, metal conduit is very robust but also labour intensive to bend, thread and fix. Tray is fairly labour intensive but more cables or circuits can be laid on a single tray compared to a conduit.
+ **How long they will last** – known as 'longevity'. Cable in conduit will last much longer than a simple cable clipped to a wall. This is because it can withstand more mechanical forces. Using a conduit, the cost of installation could be outweighed by the cost of maintaining a clipped cable as the latter could prove more expensive, with down-times and replacement costs if the clipped cable is damaged.

Down-times The duration a system is not operating due to fault or failure. Sometimes loss of supply can be very expensive for businesses.

Topic 3.3 Techniques for installing wiring components; Topic 3.4 Install wiring systems and supports

REVISED

Table 3.6 shows the commonly used wiring systems and:
+ why they are selected
+ how they are supported
+ where to find maximum support distances
+ and the maximum or minimum heights (where applicable).

75

Also shown in the table is the installation reference – this is a specially designated reference depending on how cables are installed, as detailed in BS 7671 Table 4A2. It is used to select the cross-sectional area of the cable when designing a circuit, as the method used to run the cable affects its current-carrying ability.

Table 3.6

Type of cable	Support methods	Installation method	Reasons for selection	Further information
Non-sheathed single-core	+ Conduit + Trunking	Method B	+ These cables have no protection by a sheath, so they must be installed in a containment system (e.g. conduit or trunking). + Single core cables are an advantage when systems have complex switching arrangements or need future adaptions. + Cables need to be drawn in to conduit using a draw tape.	Trunking must have a gap no greater than 1 mm when protecting non-sheathed cables.
Sheathed multi-core flat profile	+ Within a containment + Clipped direct to a surface + On cable tray + Buried in a wall + In or under thermal insulation	+ Method B + Method C + Method E + Method A + Reference 100-103	+ One of the most commonly used cables in the UK. + Universal cables that serve most types of installation. + They may need further protection by a management system, such as basket. + Using basket also means that more cables following the same route can easily be installed in some commercial installations. Might also be installed in mini-trunking for aesthetic reasons when run on the surface. + Very common in domestic installations. + Surface wiring would need dressing.	See Table D1 in the IET OSG for maximum spacings for clips when installed as Method C or 100-103. Also applies to ties or cleats for method E, where the cable is secured to the underside of a tray or a vertical tray.
Sheathed multi-core round	As flat cable above	As flat cable above	+ Used more for outdoor installations (e.g. where armouring is not required, such as cables clipped to a wall for lighting).	As flat cable above
Armoured	+ Clipped direct to a surface + On tray or ladder + Buried in the ground	+ Method C + Method E + Method D	+ A very tough cable that can stand up to many mechanical pressures – it has a steel armouring. + Suited for large supplies and outdoor or industrial uses. + Must be terminated using a gland assembly, which makes it particularly sealed and secure.	See Table D1 in the IET OSG for maximum support distances using clips, ties or cleats for method E, where the cable is secured to the underside of a tray or a vertical tray or ladder.

Drawn in Cables are pulled through conduits. This requires a lot of pre-planning to make sure the right number of cables are pulled into each section.

Draw tape Draw tape is normally nylon and rigid enough to be pushed through sections of conduit, where cables are then tied to it and pulled through by pulling the draw tape back again.

Dressing Means making sure cables are neat and straight without kinks or twists.

Check your understanding

8 Flat profile cables have installation numbers 100 to 103 rather than installation reference letters. When do reference 100 and 101 apply to these cables?

Check your understanding and progress at **www.hoddereducation.co.uk/myrevisionnotes**

Other cables include the following:

+ **Flexible:**
  + Multi-core cables which are used for short connections to machinery or appliances.
  + Often fastened by clips where required. See Table D1 in the IET OSG for distances of supports where used.
  + There are many different types, from heat resistance for heating systems, to SY flexes, which have a fine armour braiding under the outer sheath for improved impact protection.
+ **Fire retardant:**
  + These cables give good resistance against fire and are used in fire detection and alarm systems.
  + Two common types are mineral-insulated copper-clad (MICC) or fire performance (FP) cable.
  + MICC cable has the best fire performance properties as the insulation is mineral-based powder. But it is labour intensive to install compared to the FP cable.
+ **Data, signal and communication cables:**
  + There are several different types of data-comms cable.
  + Includes co-axials, fibre-optic, twisted pair (TP) and shielded twisted pair (STP).
  + Cable cores are twisted into pairs to reduce electro-magnetic interference (EMI).
  + TP and STP cables also have ratings based on their data transmission speeds and data/voice capabilities: Cat 5, Cat 5e and Cat 6. As technology improves, so will the category number.
  + Currently, fibre-optic cables offer the fastest speeds for data communications – they use light signals, rather than electrical signals which are transmitted through copper cables.
+ **Catenary wire:** This isn't technically a cable, but a method used to support cables when cables need to span across a distance (e.g. between two buildings). The catenary wire is fixed at each end and cables are tied to it so the wire takes the weight and strain, not the electrical cable. Table D2 in the IET OSG gives detail about heights of cables supported by catenary wire, including maximum lengths of span for cables.

Topic 3.2 Types of support methods and application

REVISED

Clips, cleats and ties

When cables are fastened to a surface, tray or ladder, they are supported by a clip, cleat or tie.

Ties are plastic or metal strips that wrap around the cable and tray, through the perforations in the tray, and they lock tight. They are often referred to as 'zip ties'.

Figure 3.9 shows other types of clip or cleat. It also shows a 'crampet', which is used for holding conduits in place before concrete is poured over them.

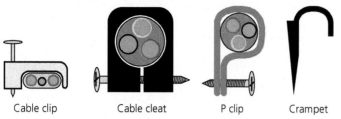

Cable clip   Cable cleat   P clip   Crampet

Figure 3.9 Types of cable support

Stopping premature collapse

BS 7671 requires all wiring systems, containment and management systems to be supported so they do not prematurely collapse in the event of a fire.

This protects firefighters who enter a burning building to tackle a fire. If a plastic support system melted during a fire, this could bring down cables and trap the firefighters and/or their breathing apparatus.

As a result, BS 7671 requires all systems to have adequate support using metallic fixings to reduce the risk of premature collapse. Metal will not melt as fast as PVC.

Many manufacturers produce metal alternatives for clips and saddles etc. which can look just the same as the plastic versions. Not all supports need to be metal – just enough to give adequate support in these situations.

> **Premature collapse** When the wiring system falls down due to fire before parts of the building structure supporting it.

Buried cables

Where cables are installed in the ground, they may be laid as follows:

+ **Buried directly in the soil**: This is the most cost-effective way to install one cable that is unlikely to need changing or updating.
+ **In service ducts**: Service ducts are like very large conduits made of either clay or PVC – usually 100 mm in diameter. These offer good mechanical protection and allow for replacement or further cables to be drawn through at a later date.
+ **In service trenches**: Trenches are normally formed of concrete and are below-ground tunnels used for electrical and other services. In trenches, cables are normally supported on cable ladders or a series of J-shaped hooks on the side of the trench.

> **Check your understanding**
>
> 11 You have been asked to install an SWA cable to supply a garden shed. The cable will run in the ground under flower beds. What will you need to do to protect the cable once in the ground? How will you warn others in the future that the cable is there if they are digging over the flower beds?

LO4 Install accessories and terminate using a range of connections

This Learning Outcome focuses on your practical skills of installing and connecting or terminating equipment, which is assessed during your synoptic assessment. In this section we will take another look at any aspects that may be in your exam.

Topic 4.1 Factors that affect the selection of accessories; Topic 4.2 Install accessories

Accessories are items which switch, control or connect appliances (or other current-using equipment) to an electrical circuit.

Most accessories are selected for the following reasons:

+ **Current rating**: How much current they are rated to switch, control or deliver.
+ **Finish**: What the accessory is made of (e.g. polished chrome, white plastic, metal-clad, brass). It could also include a style, such as Victorian polished brass.
+ **Function**: What does it do, such as switch, control, protect (does it have a fuse?), or restrict use, such as non-standard socket-outlets used for particular equipment with special plugs.
+ **Environment**: Is it suitable for its location, e.g. water-resistant, sealed, dust-proof, vapour-proof?

Table 3.7 shows the range of accessories or equipment along with the factors that affect their selection.

(Note: As current rating and suitability for the environment applies to all accessories, these are not mentioned in the table.)

> **Current-using equipment**
> A term used in BS 7671 for any appliance, load, luminaire etc. that uses current in order to function, or converts electricity into another form of energy (e.g. light or heat).
>
> **BS EN 60309** The standard for socket-outlets for varying current voltages and current ratings. They may be 16 A, 32 A, 45 A or higher. Their colour denotes their voltage rating, such as yellow for 110 V, blue for 130 V and red for 400 V three-phase.

Table 3.7

Item	Purpose	Main reasons for selection
Switchgear and control gear	+ Covers a wide range of items from distribution boards controlling circuits, isolators used to control parts of an installation or main switches for entire installations. They are used to control or switch entire installations or parts of installations.	+ Function and the number of poles they control. + Their names describe how many poles (switches) are operated by one lever. Four-pole are four switches linked to one lever; TP-N is triple pole with a removable neutral link; three-pole, double-pole or SP-N is a single pole with neutral link. + Normally in cupboards or out of sight – aesthetics and finish not a big factor.
Switches	+ Used to control or divert current. + Mainly used to control lighting but could also be used for any item of fixed current equipment too (e.g. an immersion water heater or boiler).	+ Function as above but also, where lighting is being controlled, switches may be one-way, two-way or intermediate. + As these devices are normally on show, finish is also a major factor in their selection.
Socket-outlets	+ Used to connect mobile or moveable equipment to the electrical supply. + Allow easy connection and disconnection of appliances or equipment so they can be moved.	+ Finish is one of the main reasons for selecting a socket-outlet, as these are one of the most common accessories on show in an electrical installation. + Function is also a major factor where use may be restricted or the current demand is larger than the standard 13 A outlet. + BS EN 60309 sockets, known as commando sockets, are used to connect equipment of varying current demands, as well as voltage and phases. + Protection is also a reason for selection, as standard 13 A plugs contain BS 1362 fuses, which are intended to protect the appliance and appliance flexible cable. They are commonly rated at 3 A, 5 A and 13 A.
Fused connection units (FCU)	+ Used to control and protect fixed appliances, such as hand-dryers or other items where they are not intended to move but still need the fused protection a plug provides. + Their maximum rating is 13 A. Fused connection units (or spurs) may be switched or unswitched.	+ Finish is a major factor as these are generally on show in installations. + Protection as they contain a fuse to protect the appliance and appliance supply cable. + They can accommodate any BS 1362 fuse commonly rated at 3 A, 5 A and 13 A.

Item	Purpose	Main reasons for selection
Luminaires	+ A luminaire (or light fitting) is used to provide light. + Some have interchangeable lamps and many are designed for a particular lamp. + LED are fast becoming the most common type of luminaire – they do not contain separate lamps but instead have fixed arrays of LEDs delivering the light output.	+ Finish and style are the most common reason in dwellings – aesthetics is a deciding factor as a luminaire is one of the most commonly seen electrical accessories. + Function is also important as the luminaire must deliver the correct amount and type of light. + Lighting is an important design factor in a place of work to carry out tasks safely and comfortably, or where items need to be displayed or highlighted.
Luminaire couplers	+ A luminaire coupler is the method used to connect a luminaire to the electrical circuit. + Examples include a ceiling rose which provides a fixed connection to a lamp-holder or a plug-in ceiling rose enabling the disconnection of luminaires by a small plug and socket. + Where the weight of the luminaire hangs from the coupler, these are called luminaire-supporting couplers (LSC).	+ Finish will be a factor where the coupler is on show. Some may be very decorative or bespoke, where a particular design is desired. + Many luminaire couplers are selected for their function. A plug-in type coupler enables a luminaire to be isolated for repair or replacement without the need to isolate the whole circuit, losing all lighting. + Other couplers may have four-pin plugs or connections, allowing the connection of a permanent live for emergency lighting applications.
Wiring connections and connection box	+ Connections are used to join conductors or cables. + Where possible, connections in wiring should be avoided as any connection is liable to failure or, unless designated as non-maintenance, accessible for inspection and testing. + Sometimes, wiring connections (or wiring centres) are required where complex or auxiliary circuits require control or where something has been added.	+ As many junction boxes are either out of sight or in discreet locations, finish is not normally a factor. + Function is important in terms of how many connections are needed and what type of connection is suitable. + Connections designated as maintenance-free might be concealed or buried (see Topic 4.3 for the types of connection). + The box used for connections should be made of material that is suitable for the environment and arc-resistant to stop the risk of fire should connections become loose and arcs occur.

Busbars	+ These are solid copper bars which carry a current, usually 100 A or over, and are clamped to take off a supply for a particular use such as a distribution circuit.
	+ They are housed in an enclosure on insulated supports. Four are often present for three-phase and neutral supplies.
	+ Busbar trunking is a system where the busbars are extended into locations where equipment and machines are tapped off the busbar.
	+ They can be considered as a large distribution cable but the cores are exposed along the route to take supplies from them.
		+ Normally hidden in cupboards or in voids so finish is not normally a factor.
		+ Busbars can be selected for several functions, including:
		  + distribution within switch rooms
		  + rise through a multi-storey building for each floor supply to be tapped off. These are called rising mains.
		  + run through an area, such as a workshop, where large power equipment is tapped off where needed.

Check your understanding

12 You are working in a new house build where the client has specified a modern, open-plan look and feel to the house. They ask you for suggestions for socket-outlet styles. What are the different finishes you could recommend? Suggest two that would suit this installation.

Bespoke Something that has been made for a particular client or use to a particular specification.

Check your understanding and progress at **www.hoddereducation.co.uk/myrevisionnotes**

Topic 4.3 Carry out connections

Connections bind a conductor to an accessory or another conductor. There are many different types and the choice is dictated by the manufacturer of the equipment of the accessory. Sometimes, there may be a choice, such as those used for wiring connections. These are shown in Table 3.8.

Table 3.8

Connection type	Common use	Advantages/limitations
Screw	One of the most commonly found connections within accessories. Tightening the screw clamps down on the cables or, in the case of a moving plate terminal, tightening the screw causes conductors to be squashed between two plates.	✦ Easy to use and can usually accommodate several conductors in the same terminal. ✦ Must be accessible for maintenance as they can work loose over time.
Clamp	A cable clamp can be a type of screw or nut and bolt type connection, usually used for earthing and bonding connections. The cable is clamped between two plates held, e.g. by a nut and washer.	✦ These are suitable for small conductors or, where large conductors are clamped, they will require a lug attached to the conductor by a compression tool (see 'Compression' below). ✦ Offer a good, firm connection but like screw terminals, these may work loose over time.
Compression	Using a special crimping tool, lugs are compressed or squashed onto the conductor, giving a tight and secure connection.	✦ Reliable connection which does not need to be assessable for inspection and maintenance. ✦ If different metals for lugs are used, electrolytic corrosion can occur unless special compounds are used on the cable before compressing.
Solder	Solder wire, made from tin and silver, is heated and melts and is then used to coat conductors; these can be held in place when the solder cools and sets. As solder is made of tin and silver, it offers a good conductive joint. Commonly used in printed circuit boards and fine electronic connections.	✦ Can be fiddly to make connections and not suitable for large conductors. ✦ Can provide a long-lasting durable connection.
Insulation displacement	This termination works by pushing an insulated conductor onto bladed terminals, which cut through the insulation making contact with the conductor. It is typically used for data and telecommunication cables.	✦ Very quick method of terminating conductors without the need to strip back insulation. ✦ Only suitable for very small cables.
Compact lever	Sometimes referred to as push-in connectors, the conductor, when pushed into the terminal, is locked and kept in place. Some have an accessible lever which can be used to release the cable.	✦ Very quick and reliable means of terminating a conductor. ✦ May limit the number of cables that can be terminated in one place.

> **Typical mistake**
>
> Flexible cables have very fine strands. To ensure a good connection when terminating them, a ferrule is used over the strands. What is often given as a 'good method' for terminating flexible cable is to double back the end – but this is not a good method! This is okay for solid cable but not for fine flexible cable, as stray strands may short out.

✚ Hand tools used for electrical installation tasks
✚ Power tools used for electrical installation tasks
✚ Measurement and marking tools used for setting out
✚ Factors affecting the selection of containment and management systems
✚ Types of containment and management systems
✚ Fabricating, forming and component parts of containment systems

✚ Factors affecting the selection of wiring and management systems
✚ Types of wiring systems
✚ Techniques for installing wiring systems
✚ Support methods for wiring systems
✚ Factors affecting the type of accessory selected
✚ Identifying types of accessory
✚ Techniques for installing accessories and components
✚ Types of connection used in electrical installations

Exam-style questions

1 What hand tool is used to cut a length of metal cable tray to length?
  a) Cross-cut saw
  c) Hacksaw
  b) Keyhole saw
  d) Bow saw

2 What type of driver is used to tighten the screw head shown in Figure 1?

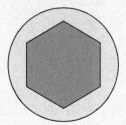

Figure 1

  a) Torx
  c) Pozi
  b) Slotted
  d) Hex

3 What tool is used to remove burrs from the inside of a metal conduit that has just been cut to length?
  a) Grips
  c) Side cutters
  b) Reamer
  d) Stillson wrench

4 What is used to accurately measure a detailed scaled line on a drawing?
  a) Measuring wheel
  c) Chalk line
  b) Laser level
  d) Rule

5 What would be the most suitable tool to mark two points at exactly the same height but in two separate rooms next to each other?
  a) Water level
  c) Chalk line
  b) Spirit level
  d) Set square

6 Which of the following is a containment system?
  a) Conduit
  c) Tray
  b) Basket
  d) Ladder

7 What is the **maximum** distance between supports for a 25 mm metallic conduit run horizontally along a wall, as given in the IET On-site Guide?
  a) 0.75 m
  c) 2.00 m
  b) 1.75 m
  d) 2.25 m

8 What is the **maximum** distance from a bend in a 50 mm × 50 mm metal trunking system, for the first support fixing along a 3 m length?

  a) 0.30 m
  c) 0.75 m
  b) 0.50 m
  d) 1.00 m

9 What type of conduit box is shown in Figure 2?

Figure 2

  a) Offset angle
  c) Tangent angle
  b) Side entry
  d) Side 90

10 What trunking feature provides segregation of circuits?
  a) Compartments
  b) Insulated PVC
  c) Suspended
  d) Galvanised finish

11 A conduit needs to contain the following cables over a length of 8 m with no bends:
  ✚ 3×2.5 mm^2
  ✚ 8×4 mm^2
  What is the **minimum** permissible diameter of conduit in accordance with the IET On-site Guide?
  a) 16 mm
  c) 25 mm
  b) 20 mm
  d) 32 mm

12 How many 25 mm^2 PVC conductors are permitted in a 100 mm × 38 mm trunking when considering the correct space factors from the IET On-site Guide?
  a) 10
  c) 20
  b) 15
  d) 25

13 Which one of the following is a suitable containment system for connecting directly to a motor?
  a) PVC trunking
  c) Flexible conduit
  b) PVC conduit
  d) Metallic conduit

14 What method of installation reference letter is given to an armoured multi-core cable clipped directly to a wall?
  a) A
  c) C
  b) B
  d) D

15 What is the **minimum** internal radius of a bend made to a non-armoured PVC multi-core copper cable having an overall diameter of 17 mm?
  a) 51 mm
  c) 94 mm
  b) 68 mm
  d) 106 mm

16 Why do wiring systems require a number of metallic supports?

   a) To stop premature collapse.

   b) To stop electrolytic corrosion.

   c) To provide a conductive earth.

   d) To provide a fault return path.

17 What is the **minimum** height for a PVC cable suspended over a road with no crossings or junctions?

   a) 3.0 m   c) 4.7 m

   b) 3.5 m   d) 5.8 m

18 Which of the following is most likely to be specified by a client regarding the selection of a socket-outlet?

   a) BS number   c) IP rating

   b) Type of finish   d) Fuse rating

19 What cable is most suitable for fire alarm systems?

   a) Thermoplastic steel-wire armoured cables

   b) Mineral-insulated copper-clad cables

   c) Flat profile thermosetting multi-core cable

   d) Braided armour multi-core flexible cable

20 What type of connection uses a tool which pushes the cable onto blades which trap the cable and makes contact with the conductor?

   a) Compact lever

   b) Moving plate screw terminal

   c) Compression

   d) Insulation displacement

4 Electrical technology (Unit 204)

This chapter examines electrical systems from generation to installation, including:

+ the basics of how electricity within electrical installation is controlled
+ how protection is provided
+ basic circuit arrangements.

It also explores the range of technical information referenced when designing, installing and maintaining systems, as well as the information provided to a client before installation work starts and during the handover process.

LO1 Understand how electricity is supplied and the characteristics of consumer's equipment

In this Learning Outcome, topics cover:

+ how electricity is generated, transmitted and distributed to the end consumer
+ how the three common types of earthing arrangements are used in relation to the supply to a consumer's installation
+ how the supply is controlled when the consumer becomes responsible for the system.

Topic 1.1 Generation, transmission and distribution of electricity

REVISED

There are three key areas for delivering an electrical supply to an installation. These are:

1 generation
2 transmission
3 distribution.

Generation

Electricity can be generated in many ways due to a large-scale move towards renewable resources compared to traditional fossil fuels, which are now in decline in the UK.

Table 4.1 shows the common methods used to generate electricity, including the basic operating principles of each one.

> **Renewable resource** A resource that can be used repeatedly and does not run out because it is naturally replaced, such as wind or solar (sunlight) energy to generate electricity.
>
> **Fossil fuels** Resources such as coal, oil or gas that are mined from the earth and burnt to produce heat – this produces carbon greenhouse gas pollution. They are not renewable resources.

Table 4.1

Fossil fuel sources	
Fossil fuels all work in the same way – fuel is burnt, which heats water to form high pressure steam. This steam turns a **turbine**, which rotates the generator at high speeds and generates electricity.	
Gas	✦ Gas is the most widely used fossil fuel. ✦ It can produce instant heat, so a gas generator can heat water to steam much faster than coal generators. ✦ Its use is in decline due to the need to reduce greenhouse gases produced when gas is burnt. ✦ But it remains the largest source of electricity generation in the UK in 2021.
Oil	✦ Oil/diesel is used for many regional generators, which power local areas at times of **peak demand**. ✦ The generators can deliver electricity immediately. ✦ The generators are used for private stand-by generators for buildings that need power supplies during power cuts (e.g. hospitals have back-up generators in case of power failures).
Coal	✦ Coal was the most widely used fuel used to produce high pressure steam for the turbines. ✦ It is highly polluting so has been scaled down in the UK, with only a few coal-fired power stations remaining. Most of these are only used during periods of peak demand. ✦ In 2017, the UK went 24 hours without using a single coal-fired power station for the first time since 1882. In 2020, the entire summer months went without coal being burnt. This is a huge step forward in reducing pollution and climate change.
Renewable resources	
Wind	✦ Wind power turns a propeller which directly drives a generator. ✦ Wind is a natural resource, so no pollution or resources are required after production of the turbine. ✦ The use of wind energy is ideal for the UK – it is an island with a lot of wind from the sea. ✦ The use of wind energy is rising each year in the UK, with many offshore wind farms being constructed several miles out to sea.
Wave	✦ The sea produces a huge amount of energy, either in the form of waves or as the tide moves in and out each day. ✦ This movement of water is used to rotate generators through a water-wheel effect. ✦ Electricity is also produced by using waves in a back-and-forth motion rather than a rotating one.

Hydro	✦ There are three types of hydro generator:
	  ✦ Water courses that are held by a dam and released through pipes diverting the water through turbines.
	  ✦ Flowing rivers where a **weir** across the river harnesses the water's energy.
	  ✦ Stored hydro generators where a large volume of water is held in a reservoir high up on hills or mountains.
	✦ At times when instant electricity is needed due to high demand, the water is released from the reservoir through turbines that generate electricity.
	✦ When the station is off-line, the water is pumped back up to the reservoir ready for the next time it is needed.

Photo-voltaic	✦ A photo-voltaic (PV) cell uses a combination of materials, which convert solar energy into electricity without any moving parts. You can see these on the roofs of some houses or in fields called solar farms.

Turbine A turbine is basically a series of propellers which are turned at very high speeds by high-pressure steam or water.

Peak demand When more electricity is being used around the whole country, such as cold, dark days when more heat and light are needed. Sometimes, this can be at half time during a major football game on TV, when everyone boils a kettle to make a cup of tea!

Weir A low dam across a river that increases the force of the water as it flows over the top. Sections of a weir can be raised or lowered to regulate the force of the water.

Typical mistake

Stored hydro is where the water used to turn the turbines is stored until needed – it isn't to store the electricity generated in batteries.

1   Biofuel is fast becoming an alternative fuel used to generate electricity, as well
    as a fuel source for heating boilers in offices and houses. What is biofuel and is it
    carbon neutral?

Transmission systems

In the UK, the transmission system is called the national grid. It is a network
of cables, mainly over ground, which is used to send electricity all around the
UK from the generator stations.

As the transmission systems are hundreds of miles long and send vast
amounts of electrical power around the UK, the systems use a range of very
high voltages. These are:

+ 400 kV (known as the super grid)
+ 275 kV
+ 132 kV.

There are two main reasons high voltages are used:

1   High voltages mean reduced current – so smaller conductor sizes can be
    used.
2   As the cables travel vast distances, the voltage lost due to cable resistance
    has less of an impact on high voltages than it would on lower voltages.

A locality consumes 80 MW of electricity. What would be the current demand at
400 kV and at 400 V?

To calculate the current demand based on power and voltage values, use:

$$I = \frac{P}{V}$$

So, at 400 kV and remembering that 80 MW is 80×10^6 W:

$$\frac{80 \times 10^6}{400,000} = 200 \text{ A}$$

So, the cables need to be big enough to carry 200 A to the area of demand.

To calculate the current demand at 400 V, the same process is used, so:

$$\frac{80 \times 10^6}{400} = 200,000 \text{ A or } 200 \text{ kA}$$

At this level of current, the cables would need to be huge! So, the benefit of increasing
the voltage is to reduce the current values, making cable sizes much more realistic.

1   If a section of a transmission system had a resistance of 30 Ω and supplied a
    locality having a demand of 25 MW, what would be the voltage loss at 400 kV and
    132 kV?

The issues with transmitting at higher voltages are as follows:

+ **High voltages can break down insulation:**
    + When cables are over ground, air is used as an insulator between
      conductors.
    + In underground cables, materials such as PVC are used as insulators
      between conductors.
+ **Conductors need to be suspended high above ground and several metres
  away from each other:**
    + High voltages can jump across or break down the air resistance,
      especially when the air has a high water content (i.e. relative humidity).
    + The towers used to do this are known as pylons.

- The higher the voltage used in the transmission system, the bigger the pylon must be.
- Pylons used for the 400 kV super-grid are as high as a tower block and have six cables suspended from them (three on each side).
- There is a single cable running from the top of the pylon linking each pylon which acts as a common earth.

Where voltages change within the transmission system, transformers are used to do this by either stepping up the voltage (e.g. 132 kV to 400 kV) or stepping down.

Where transmission systems enter urban areas or pass through the countryside, they may need to run underground. In this case, the voltage will be stepped down (e.g. 132 kV) so it will not damage the insulation between the conductors in the underground cable.

Distribution systems

At points around the national grid, electricity is tapped off to be distributed to the user. These systems are known as the distribution systems. They are looked after by the distribution network operators (DNOs).

As distribution is much more localised, the voltages can be stepped down to lower values – this keeps the pylon sizes smaller and less of an eyesore!

Also, lower voltages mean cables in urban areas can be run underground, keeping the supplies invisible. As underground systems are more expensive, rural supplies will normally use cheaper overhead supplies.

Depending on how far the electricity needs to be distributed, distribution voltages may be:
- 33 kV
- 11 kV
- 400 V
- 230 V.

In most cases, the underground cables in towns and cities are 11 kV. These supply the many sub-station transformers.

On the outgoing side of the sub-station to consumer installations, the supplies are 400 V three-phase (or 230 V single-phase depending on their overall current demand). As a general rule:
- buildings having a demand over 100 A will have a three-phase supply
- buildings having a demand below 100 A will have a single-phase supply.

Large buildings and facilities may have their own sub-station transformer. These places may be supplied at 11 kV or 33 kV.

> **Exam tip**
>
> In your exam, questions can be related to understanding from other units in one question. As well as knowing whether step-up or step-down transformers are used, you may also be asked in the exam how transformers work, including their ratios. Make sure you revise those principles in Chapter 2 of this book.

> **Sub-station** The final transformer before the consumer – they are found in many different places, such as behind fences, at the end of roads or inside brick buildings.

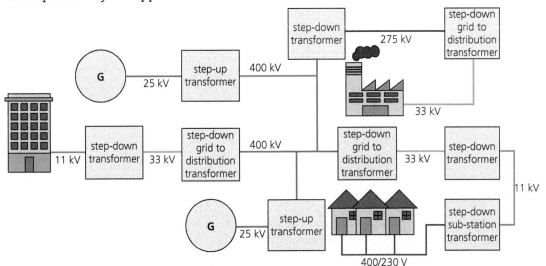

Figure 4.1 The generation, transmission and distribution system, where G is the generator station

My Revision Notes: City & Guilds Level 2 Advanced Technical Diploma in Electrical Installation (8202-20)

Topic 1.2 Electrical intake arrangements

In this topic, the focus is on the electricity supply to a standard single-phase 230 V consumer installation.

Figure 4.2 shows the typical layout for a supply arrangement within a consumer installation.

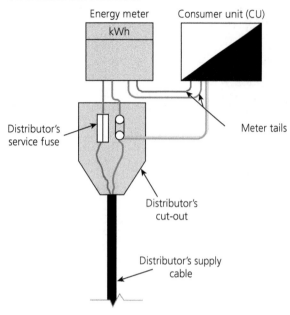

Figure 4.2 A typical single-phase supply arrangement

Table 4.2 lists the items of equipment in a standard arrangement.

Table 4.2

Distributor's cut-out	✛ This enables the supply DNO's cable to be terminated and the supply fused by the service fuse. ✛ Service fuses are typically 63 A or 80 A rating. ✛ Cut-outs and service fuses are sealed by the DNO to stop people from tampering with the electricity supply. ✛ The cut-out is where the DNO's responsibility ends.
Electricity energy meter	✛ The meter is used to measure the quantity of energy used. ✛ The meter is owned by the supplier of electricity to the installation (may not be the DNO). ✛ The energy supplier is chosen by the consumer and is responsible for the meter and tails on the ingoing side.
Meter tails	✛ The meter tails are the cables that enter and leave the energy meter. ✛ Where the meter tails link the energy meter directly to a consumer unit (CU), they are the responsibility of the consumer. ✛ Meter tails should be 25 mm² insulated and sheathed single core cables with copper conductors.
Electricity isolator switch	✛ In situations where the meter is remote from the consumer unit, the energy supplier or DNO is likely to install an isolator switch directly after the energy meter. ✛ This allows the cable to the CU to be isolated and protected by a protective device (other than the DNO service fuse). ✛ Not all installations will have these isolators if the consumer unit is close to the energy meter.

Many electrical installation supply arrangements differ depending on where the cut-out and meter are located in relation to the consumer unit.

Over many years, new properties were fitted with meter cupboards located on the **outside** of the property, allowing the meter to be read by the supplier without the need for the occupants to be at home.

Check your understanding and progress at **www.hoddereducation.co.uk/myrevisionnotes**

In these situations, it is likely an isolating switch is installed after the meter, as the consumer unit is located more than two or three metres away – this means the cable to the consumer unit is protected and easily isolated.

With smart metering offering readings over the internet, the position of the meter is more likely to be where you can access Wi-Fi.

> **Check your understanding**
>
> 2   Figure 4.2 shows a typical intake arrangement. What type of earthing arrangement is shown and why?

Topic 1.3 Features of consumer units/ distribution boards

REVISED

What is the difference between a consumer unit and a distribution board? They are basically the same thing – but there is an unwritten rule that a consumer unit (CU) is the first point of distribution in an installation, especially a domestic type of installation. Any further boards are called distribution boards (DB).

For example, if a house had a consumer unit next to the meter, but one of the circuits from the CU went into the garage where there was a further board, the garage board would be classed as a remote DB.

Figure 4.3 shows the features of a consumer unit which is classed as a split-way board.

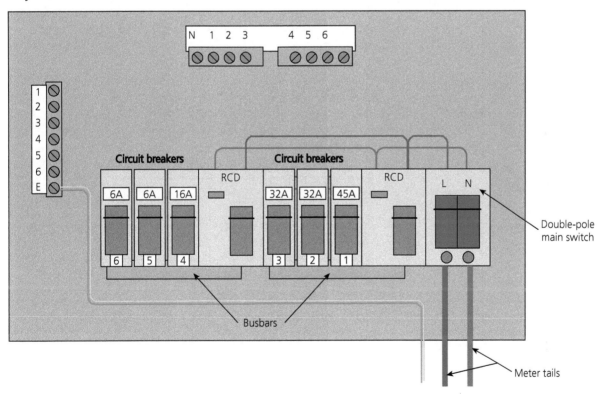

Figure 4.3 A split-way type consumer unit

> **Now test yourself**
>
> TESTED
>
> 2   You are looking to purchase a distribution board for a new job starting in a few days. The DB needs to accommodate six circuits. You find a modular board where you need to purchase two RCDs and a main switch unit to form a split load board. How many 'ways' and 'modules' does the DB need to be to accommodate all this?

In the split-way board, the RCDs each protect a number of circuits providing the necessary **additional protection** (as required by BS 7671).
+ The RCD supplies the circuit breakers via the solid copper busbar which connects into the supply side of the circuit breaker.
+ Split load boards are arranged this way because RCDs are sensitive devices that can trip easily. If this happens, circuits protected by the other RCD will still work.
+ A split-way CU is a cost-effective way of providing RCD protection across all circuits without using more expensive RCBOs (residual circuit breaker with overload).

Standard CUs that are **not** split way will just have a main switch and use individual RCBOs to protect each circuit. RCBOs are circuit breakers with a combined RCD protection function.

Exam tip

Section 3 of the On-site Guide (OSG) contains some different intake and consumer unit arrangements. You can refer to this in the exam to help you, so be familiar with the information contained in the OSG.

Check your understanding

3 Look at the split load board arrangement in Figure 4.3. Imagine you are connecting the six circuits controlled from the CU. By mistake, you connect the neutral of circuit 3 into the number 4 neutral connection, and the circuit 4 neutral cable into neutral connection 3. What would happen if the installations were powered up?

Consumer unit features for domestic and small commercial, single-phase installations are shown in Table 4.3.

Design current The full load current of the circuit under normal operating conditions.

Table 4.3

Main switch (double-pole isolators)	+ Every domestic electrical installation must have a double-pole linked main switch to isolate the whole installation with one switching action. + Some small commercial installations do not always need a double-pole main switch – but it would always be advisable and easier to install one. + They must be rated as switches (even though they are called isolators), meaning they can switch the full load current.
Overload and fault protection	+ Protection against overload and faults is provided by the circuit breakers (or other protective devices; see Topic 2.1) that protect and control each individual circuit. + The circuit breaker nominal rating (I_n) must co-ordinate with the circuit. + It must be equal to, or higher than, the circuit **design current** (I_b). $$I_n \geq I_b$$

Additional protection against electric shock	+ BS 7671 requires nearly every circuit in a domestic type installation to have additional protection by an RCD.
	+ The protection is in addition to the basic protection by insulation in case insulation or barriers become damaged, exposing hazardous live parts.
	+ The RCD is a very sensitive device that trips at very low current imbalance between line and neutral.
	+ When providing additional protection, BS 7671 requires the residual tripping current to be no more than 30 mA (milli-amps) – it means the RCD should trip very quickly if 30 mA or more current went through the line conductor, compared to the neutral, as some is going to earth.
	+ Additional protection is needed in situations such as those listed below:
	  + **Socket-outlet circuits**: where any appliance in very poor condition could be plugged in. If the appliance supply cable of the plug was damaged, a person could easily touch a live conductor. Some cheap appliances may be risky and not meet safety standards.
	  + **Mobile equipment used outdoors**: (e.g. lawn-mowers, hedge trimmers etc.) are high risk as they use long extension leads – these can be cut by machinery or scraped around brick walls etc., leaving bare live conductors.

Check your understanding and progress at **www.hoddereducation.co.uk/myrevisionnotes**

	+ **Cables concealed in a wall**: are at risk of having their insulation damaged and penetrated by nails and screws, leaving the nail or screw live to touch. + **Circuits with luminaires**: modern luminaires have little space for the connection of the cables – this means insulation can be pinched or trapped, causing a shock risk. Luminaires can also be bought online, so they may not meet safety regulations, which is a risk. + **Circuits supplying equipment in a bathroom**: bathrooms are 'special locations' due to the higher risk of shock (e.g. water splashing on electrical equipment/getting through barriers and enclosures). Also, lack of clothing/humid environments lower body resistance, which means lower voltages can cause harm.
Consumer unit assembly material	+ Consumer units installed within domestic installations must have an outer case made of fire-resistant material, such as metal. + This reduces the spread of fire or heat accumulation if there is a fault in the unit due to loose terminals or connections.

Typical mistake

Many think a switch and an isolator are the same thing – but they're not! A switch is designed to operate under full load conditions, while an isolator (if frequently used to switch a full load current) will deteriorate and burn out the connections.

Exam tip

Remember, the maximum residual current rating for an RCD providing additional protection is 30 mA. This is a fact you are expected to know, and although you can look it up in the OSG, remembering it will save a lot of time.

Check your understanding

4 With few exceptions, all circuits in a domestic installation require RCD protection. So, why do we not protect all circuits with one single RCD as a cost-effective way of providing the RCD protection?

5 Can you think of any circuit in a domestic installation that does not need additional protection by an RCD?

Topic 1.4 Types of earthing arrangements

REVISED ○

In the UK, there are three types of earthing arrangement that can be found within an electrical installation.

The earthing arrangement uses letters to designate how the earth and neutral are connected in the supply and installation. These are listed below:
1 **TN-S** where the earth (T) and neutral (N) are separate (S) in the supply and installation (see Figure 4.4).
2 **TN-C-S** where earth (T) and neutral (N) are combined (C) in the supply but separate (S) in the installation (see Figure 4.5).
3 **TT** where there is no link to neutral, but the supply sub-station has an earth electrode (T) and the consumer's installation has an earth electrode (T) (see Figure 4.6).

The earthing arrangement is very important when disconnecting a circuit under fault conditions to protect against the risk of electric shock.

This is because the earth path resistance will change, depending on what is used as the earth return path.
+ An increase in resistance will cause a decrease in fault current.
+ A decrease in fault current will increase disconnection times, creating a higher risk.

Earth Also known as 'terra-firma' from the French word for 'earth' –'terre'. Many international electrical regulations are written in French, so 'T' is the symbol for 'earth' when looking at earthing arrangements.

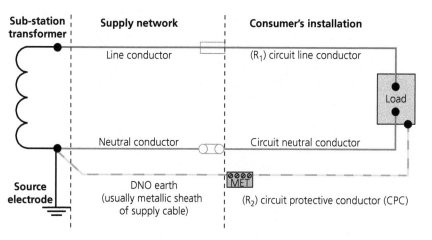

Earth fault loop path for a TN-S earthing arrangement

Figure 4.4 A TN-S earthing arrangement

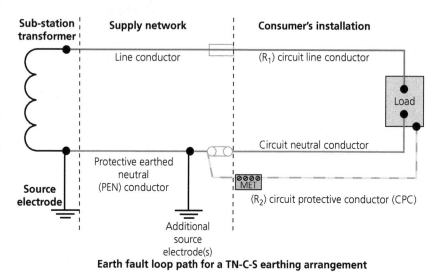

Earth fault loop path for a TN-C-S earthing arrangement

Figure 4.5 A TN-C-S earthing arrangement

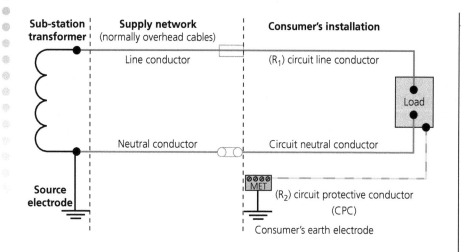

Earth fault loop path for a TT earthing arrangement

Figure 4.6 A TT earthing arrangement

Check your understanding and progress at www.hoddereducation.co.uk/myrevisionnotes

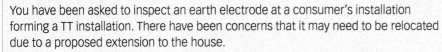

Now test yourself | TESTED ○

You have been asked to inspect an earth electrode at a consumer's installation forming a TT installation. There have been concerns that it may need to be relocated due to a proposed extension to the house.

Research the following from the IET On-site Guide regarding earth electrodes:

3 What can be used as an earth electrode?

4 What is the recommended maximum earth fault loop impedance of an arrangement, using an earth electrode as part of the earth fault path?

5 How can the weather affect the earth fault path?

> **Exam tip**
>
> There are two further earthing arrangements in BS 7671 that you may see featuring in your exam as the wrong answers, known as 'distractors'. These systems are:
> + TN-C: not permitted in the UK
> + IT: where the system has an intentional high impedance earth, which is used for very specialist installations such as some medical locations.

LO2 Understand isolation and protection

This Learning Outcome looks at the devices within an installation used for protection, control, isolation and switching, where they are installed and their basic features.

Topic 2.1 Types of protection devices   REVISED ○

Protective devices are installed to protect circuits against the following problems:
+ **Earth faults:**
    + Faults where line conductors come into contact with earth due to a breakdown in insulation.
    + This causes a high current flow, making exposed conductive parts become live to a dangerous voltage.
    + This increases the risk of electric shock.
+ **Short circuits:**
    + Faults where line and neutral conductors come into contact due to a breakdown in insulation.
    + This causes a high current flow.
    + This increases the risk of fire.
+ **Overloads:**
    + Where too much current is drawn into the circuit by connecting the current using equipment.
    + This leads to more current than the circuit was designed for, which will make the circuit cables too hot.
    + It leads to insulation breakdown.
+ **Additional protection:**
    + Where the circuit basic protection fails due to the environment, such as an extension lead supplying outdoor equipment becoming snagged.
    + The outer insulation can be damaged, leaving a bare live conductor.
    + This increases the risk of electric shock through contact.

> **Exposed conductive parts** These are metal parts that have electrical parts but would not normally be live; but they could become live if a fault happens (e.g. metal casings of appliances or metal trunking and conduit).
>
> **Basic protection** A technical term used in BS 7671 – refers to the insulation around live parts, or the barriers and enclosures housing live parts, which prevent users from touching live parts.

There are three main groups of protective devices:

1 **Fuses:**
    + Have a fuse wire element which heats up with current.
    + If the current steadily reaches high values due to overloads, the wire melts over a period of time. The circuit disconnects.
    + If the current suddenly reaches high values due to a fault, the wire blows very quickly. The circuit disconnects.

2 **Circuit breakers:**
    + Have a magnetic coil – when a fault current reaches a pre-set value, the magnetic field causes mechanical movement.
    + This causes the switch to rapidly trip, disconnecting the circuit.

93

+ They also have a thermal trip (normally a bi-metallic strip) which causes gradual mechanical movement when heated by lower overload current.
+ This also causes the switch to rapidly trip, disconnecting the circuit.

3 **Residual current devices (RCDs):**
+ Have either electronic devices or a small toroidal transformer – it monitors current entering the circuit through the line, and back on the neutral.
+ If the circuit is healthy, they remain identical.
+ If a small fault happens, current flows to earth; the line has more current than the neutral.
+ If the imbalance current exceeds the device residual current setting, the device trips instantly. The most common residual current setting for an RCD is 30 mA or 0.03 A so these devices are highly sensitive.

Devices also have different ratings, which are:

+ **Breaking capacity (kA):**
  + The amount of current a fuse can take before it blows up or seriously damages the carrier and surroundings.
  + In the case of a circuit breaker or RCBO, faults higher than their breaking capacity could also weld together the contacts so it wouldn't cut off the current.
+ **Nominal rating (I_n):**
  + The current value that the device can handle through it in continued service (e.g. a 32 A circuit breaker can carry 32 A load current for the lifetime specified by the manufacturer, without breaking or deteriorating).
+ **Activation current (I_a):**
  + The amount of current needed to disconnect the device in the time required by the type of circuit (e.g. a socket-outlet circuit on a TN system must disconnect in 0.4 seconds).

There are several types of protective device used to protect circuits. Each one has different characteristics, which are outlined in Table 4.4.

Bi-metallic strip Two different metals bound together. Each metal expands at a different rate, which causes the strip to gradually bend with heat. The bending movement is used to activate a switch.

Toroidal Circular or doughnut-shaped.

Table 4.4

Device BS number and name	Characteristics	Uses

BS 88-2 and BS 88-6 high rupturing capacity (HRC) or high breaking capacity (HBC) fuses
 	
(BS 88-6 now discontinued)
 	
Figure 4.7 HRC or HBC fuse
	+ One or several fuse wire elements are contained in a glass barrel with sand.
	+ The heat generated causes the sand to turn into glass, absorbing the blast energy. It stops the whole fuse body exploding.
	+ By having multiple elements, this creates several smaller blasts as each wire erupts.
	+ It makes them suitable for very high fault currents, which are likely within large installations having high current supplies.
	+ They have metal tags on each end, which are bolted into position – they are very sturdy.
	+ Available in two types for disconnection:
	  + gG for general use and quicker disconnection
	  + gM for motor applications (but less sensitive).
		+ Can only be maintained by skilled persons, so unsuitable for domestic use.
		+ Able to break very high fault currents (up to 80 kA).
		+ Provide fault and overload protection.
		+ Disconnection time relative to fault current – the lower the current, the longer the time to disconnect.
		+ Large range of ratings up to several hundreds of amps.

Check your understanding and progress at **www.hoddereducation.co.uk/myrevisionnotes**

Device BS number and name	Characteristics	Uses
BS 88-3 Cartridge fuses **Figure 4.8** Cartridge fuse	+ A single fuse wire element is contained in a glass barrel with sand. + The heat generated causes the sand to turn into glass, absorbing the blast energy. It stops the whole fuse body exploding. + This makes them suitable for mid to high fault currents. + The barrel has no tags, so the fuse clips into position.	+ Suitable for any type of installation. + Provides fault and overload protection. + Able to break currents (up to 16 kA). + Disconnection times relative to current. + Good range of ratings (up to 100 A).
BS 3036 Rewireable fuses **Figure 4.9** Rewireable fuse	+ A single fuse wire suspended between two screws in the holder. + Blast containment is minimal. + It is used for asbestos (which still may be present on older carriers).	+ Difficult to replace wire when blown – not user friendly. + Provides fault and overload protection. + Able to break currents (around 1 kA to 4 kA) depending on carrier. + Disconnection times relative to current but can change due to being exposed to the air. + Minimal range of ratings.

BS EN 60898 Circuit breakers	+ These have a magnetic trip for fault currents and a thermal trip for overloads.
	+ Two body styles:
	  + Miniature circuit breakers (MCB) for lower current circuits up to 125 A.
	  + Moulded case circuit breakers (MCCB) for larger circuits where fault currents can be much higher.
	+ The case type affects the ability to disconnect at high fault currents by containing the blast.
	+ There are three different types:
	  + **Type B**: the most sensitive where the magnetic trip is set to 5 times the current rating (I_n).
	  + **Type C**: less sensitive, where the magnetic trip is set to 10 times the current rating (I_n). Makes them suitable for loads having large start up currents (e.g. motors).
	  + **Type D**: least sensitive, where the magnetic trip is set to 20 times the current rating (I_n). Makes them suitable for loads having very large start up currents (e.g. large power motors or machines).
		+ User-friendly – no parts to replace, just reset if tripped.
		+ Provide fault and overload protection.
		+ Able to break currents up to 10 kA (MCB) or higher (MCCB), depending on manufacturer.
		+ Contacts can weld together if fault currents are too high – disconnection will not happen.
		+ Good range of ratings.
		+ Different types to suit different circuit applications.

Device BS number and name	Characteristics	Uses
Residual current devices (RCD)	+ These monitor the imbalance of current between line and neutral. + They disconnect at a very low current imbalance. + Residual current settings can range from 10 mA to 20,000 mA (20 A) – but mainly 30 mA is used. + These devices need to function at very low energy currents – so the mechanisms may stick and not release at low current energy values. + The devices have a test button that releases the mechanism – this stops it from sticking if regularly used.	+ User-friendly – no parts to replace, just reset if tripped. + 30 mA devices provide additional protection. + Require maintenance by regularly pressing the test button to remain safe. + Provides very good electric shock protection from low earth faults. + Does not provide short circuit or overload protection.
BS EN 61009 Residual circuit breaker with overload (RCBO)	+ These are a miniature circuit breaker and RCD in the same body. + They have the characteristics of both types of device.	+ User-friendly, just reset if tripped. + Provides fault protection, additional protection and overload protection. + Have all the characteristics of a circuit breaker and RCD.

Topic 2.2 Purpose of discrimination/ selectivity devices

REVISED

Circuits should be co-ordinated to reduce the risk of danger or inconvenience. This means if a fault happens, the device protecting that circuit disconnects – **not** a larger device closer to the supply of the installation. Otherwise, this would cause more circuits to be lost.

Ensuring that the local device disconnects is called 'selectivity', which replaced the term 'discrimination' in the 18th edition of BS 7671 (2018).

Because different types and ratings of protective device disconnect at different current values, it isn't a guarantee that one device rated at a lower current than another device will disconnect first.

To understand this, the time current characteristics of a device need to be introduced.

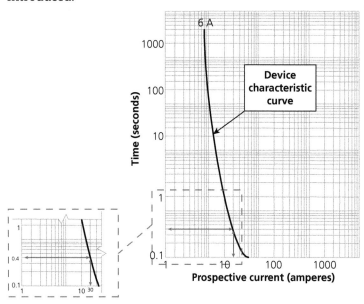

Figure 4.10 Time current characteristic graph for a 6 A BS 88-2 fuse

Figure 4.10 shows the disconnection characteristics for a 6 A BS 88-2 HRC fuse. Using the graph, you can see that a fault current of approximately 28 A would disconnect the fuse in 0.4 seconds.

When choosing devices for selectivity, you should:
+ compare the graphs for the two different devices
+ and check that the two characteristic lines do not cross each other.

If they do at any point, the higher rated device may disconnect before the lower one.

Figure 4.11 shows two situations:
1 Clear selectivity between a 32 A Type B circuit breaker and a 63 A BS 88-2 fuse.
2 Poor selectivity where the characteristics for a 32 A Type D circuit breaker cross those of the 63 A BS 88-2 fuse. This means the main fuse will disconnect first at higher currents.

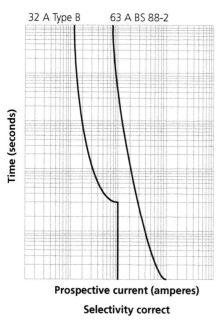

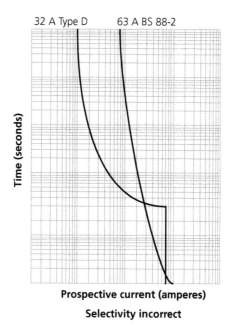

Figure 4.11 Correct selectivity and incorrect selectivity

Check your understanding

6 Looking at Figure 4.11, why does the circuit breaker characteristic line have a curve and straight line while the fuse only has a curve?

Topic 2.3 Purpose of isolation and switching

REVISED

A typical electrical installation will contain many different switches or isolators. It is very important to know what the purpose of the device is to select the right type of device.

BS 7671 requires isolation and switching devices for the following reasons:
+ **Isolation:**
    + a means of cutting off electricity to all or part of an installation
    + allows safe electrical work to be carried out.
+ **Switching for mechanical maintenance:**
    + a means of safely stopping a machine, equipment or appliance from operating in any way
    + allows safe non-electrical work which does not involve exposure to potentially live parts.
+ **Emergency switching:**
    + a means of quickly stopping a machine, equipment or appliance from operating in the event of an emergency.

+ **Functional switching**:
  + a means of controlling a machine, equipment or appliance, such as turning it on or off.

Isolation

+ Isolation must be provided for all installations in the form of a main switch.
+ The type of installation will affect the type of isolator used. Dwellings and small commercial single-phase installations will use a double-pole main switch.
+ Isolation may also be achieved for individual circuits by the protective device for the circuit.
+ In all situations, the device used for isolation must be capable of being secured in the open position. For main switches and circuit breakers, this is normally achieved with a padlock and a special locking device (if needed).
+ Where circuits are protected by fuses, the fuse must be capable of being fully removed. It should be kept by the person who undertakes isolation.
+ Devices are available which block off the fuseway – this means other fuses cannot be inserted while the circuit is securely isolated.
+ The circuit should always be isolated at the start of the circuit whenever an electrically skilled person works on it.

Switching for mechanical maintenance

Mechanical maintenance covers many tasks to a variety of different items of equipment. Providing that the work does not involve exposure to electrical terminals, it can be done by (electrically) unskilled persons. This work could include:

+ cleaning filters in an extraction system
+ installing or altering the pipework to a shower
+ replacing lamps in a fluorescent luminaire
+ servicing or cleaning an oven.

For these situations, a device capable of switching full load current must be located 'local' to the equipment. Ideally, the device must be in a position where the person carrying out the mechanical maintenance can keep the switch under their effective supervision so nobody else can switch it on.

Common devices for this purpose include:

+ fused spur connection units
+ double-pole switches
+ plugs and socket-outlets.

If the device cannot be easily supervised then the device should be lockable (e.g. a lockable rotary switch). This removes the risk of someone else switching it back on.

Emergency switching

Emergency switching devices must be capable of cutting off full load current. In domestic dwellings, the main switch can act as an emergency switch – which is normally coloured red.

If a resident within a dwelling had concerns about the safety of an electrical system, such as a major water leakage, they are able to cut off the supply easily.

In other installations where items (e.g. rotating machines) can cause a danger, emergency switching is normally provided by stop buttons.

These should be located near to the danger so it can easily be operated by a person in danger.

Check your understanding and progress at **www.hoddereducation.co.uk/myrevisionnotes**

Functional switching

This is the most common type of switch found in an installation. It allows equipment to be switched on or off as needed. It should be in a convenient place. An example would be a light switch for a room, located by the entrance door to that room.

Functional switches may also be:
+ timeclocks
+ PIR sensors
+ plugs and socket-outlets (rated at or below 32 A)
+ switches on socket-outlets
+ contactors
+ standard switches
+ push switches
+ switched fused connection units.

Devices that should never be used as a functional switch include:
+ fuses
+ luminaire connections such as plug-in ceiling roses
+ un-switched fused connection units
+ socket-outlets rated above 32 A.

LO3 Understand automatic disconnection of supply

Automatic disconnection of supply (ADS) is the most common protective measure against electric shock used in electrical installations.

It works by using a network of protective conductors, protective devices and basic protection.

The purpose is to achieve a low loop impedance for the fault current, because Ohm's Law plays a big part in it being effective as:
+ a low impedance creates
+ a high fault current, which enables
+ quick disconnection of protective devices.

In this Learning Outcome, we will revise what is needed to create effective ADS.

Topic 3.1 Principles of basic protection

REVISED ○

Although ADS is technically protection where, if a fault happens, it will make it safe, installations need basic protection.

Basic protection is intended to stop contact with parts that are normally live. This is done in two ways, as shown in Table 4.5.

Table 4.5

Insulation of live parts	+ Some live parts that do need to be accessed for terminations are covered with insulation. + This includes items such as conductors in cables or some live parts in equipment and accessories. + Insulation is defined as a part that is removed by destruction. + Insulation must be suitable for the intended voltage. + Most insulation, such as thermoplastic PVC used on standard cable, is rated around 300 to 500 V (depends on the manufacturer).

99

Barriers and enclosures	+ Barriers and enclosures are intended to house live parts that need to be accessible (e.g. terminals on a socket-outlet) to enable the cables to be connected.
	+ There are several rules that BS 7671 requires for barriers and enclosures. These are as follows:
	    + Live parts are only accessible by the use of a tool or key so access is intentional not accidental.
	    + All surfaces of a barrier or enclosure must have a level of IP protection no bigger than IP2X or IPXXB, with the exception of the horizontal top surface cited below.
	    + Any accessible horizontal top surface must have a degree of IP protection no less than IP4X.

IP protection An international numerical code system for the levels of protection against the ingress of dust or foreign bodies (represented by the first number) or water and moisture (represented by the second number).

IP2X The barrier or enclosure has no hole bigger than 12.5 mm in diameter, so fingers cannot access the inside of a barrier or enclosure. The 'X' means that protection against water or moisture is not applicable in this situation.

IPXXB A specific electrical code which means that if a finger enters the enclosure, any live parts are spaced more than 80 mm from the hole, so a finger length could not touch the live part.

IP4X A specific electrical code which means there is no hole with a diameter greater than 1 mm.

In some situations where there is a greater risk of failure of basic protection due to the location or environment, circuits need to have additional protection by an RCD or RCBO having a residual current setting no more than 30 mA.

(See Topic 1.3 in this unit for greater detail regarding environment and location.) As a reminder, situations include:
+ socket-outlets rated up to 32 A
+ mobile equipment rated up to 32 A used outdoors
+ cables concealed in a wall to a depth less than 50 mm and having no metallic earth protection around them
+ circuits with luminaires
+ circuits in a location containing a bath or shower.

Topic 3.2 Principles of fault protection

REVISED ○

If an earth fault happens in a circuit, accessory or current-using equipment, there is a great risk of electric shock if someone touches something metal that has become live.

Circuits need to disconnect quickly to protect someone in the event of a fault and remove the risk of danger.

If a high fault current continues without disconnecting, cables and equipment will be subjected to excessive current – this causes them to deteriorate very quickly due to the heat the current causes. In some cases, this could result in a fire.

Disconnection for circuits depends on three factors:
1 the circuit type
2 the circuit rating
3 the earthing arrangement.

Table 4.6 shows the maximum permitted disconnection times (as given in BS 7671) for the different types of circuit and earthing arrangements.

Table 4.6

System	Circuit type	Rating	Disconnection time (seconds)
TN-S or TN-C-S	Distribution	Any	5.0
	Final circuit supplying fixed equipment	> 32 A	5.0
		≤ 32 A	0.4
	Final circuit supplying socket-outlets	> 63 A	5.0
		≤ 63 A	0.4
TT	Distribution	Any	5.0
	Final circuit supplying fixed equipment	> 32 A	1.0
		≤ 32 A	0.2
	Final circuit supplying socket-outlets	> 63 A	1.0
		≤ 63 A	0.2

Topic 3.3 Purpose of earthing and bonding

Disconnection will only happen in the time required if the protective devices are subjected to a high enough fault current.

A high enough fault current will only happen if the fault path has a low resistance, so the network of protective conductors is important to give this.

There are two types of protective conductor:
1 earthing
2 bonding.

Earthing

The purpose of earthing is to provide a low resistance or impedance path to the means of earthing and to carry the fault current to that means of earthing.

The two common types of earth conductor are:
1 **Earthing conductor**
   + This links the means of earthing to the installation main earthing terminal (MET).
   + It needs to be capable of carrying the largest earth fault current.
2 **Circuit protective conductor (CPC)**
   + The CPC links the MET to all points in the circuit.
   + Even if there are no metal parts that need to be connected to earth, a CPC is still always run to each point.
   + The CPC needs to have a low resistance, known as (R_2), and be capable of carrying the earth fault current for the circuit.

Bonding

The purpose of bonding is to ensure any extraneous conductive parts are connected to the electrical earth. If any electrically earthed equipment becomes live due to a fault, so does the extraneous part. This means there is less risk of an electric shock because of equal potentials.

There are two bonding conductors, shown in Table 4.7.

Means of earthing The means of earthing for an installation is the earth provided by either the DNO, for a TN arrangement, or the consumer's earth electrode for a TT arrangement.

Extraneous conductive parts Metal parts that have a potential path to earth but are not part of the electrical system, such as metal gas pipes.

Equal potential Where two parts have the same voltage, which actually reduces the risk of shock. If two surfaces were live to 230 V and someone touched them both, the voltage is the same across the person, so the potential difference is 0 V and harmful current will not flow.

Typical mistake

You need to know the key difference between earthing and bonding. They are both very different applications but as they both use green and yellow cables, many people think they are the same thing!

Table 4.7

Main protective bonding conductor	This conductor links the MET to the following extraneous parts: + Metallic gas installation pipes within 600 mm of their point of entry or 600 mm on the consumer's side of the meter. + Metallic water installation pipes within 600 mm of their point of entry or 600 mm on the consumer's side of the stopcock. + Other metal service installation pipes, such as oil. + Metallic structural parts of a building which are exposed to touch. It must have a potential path to earth to be considered extraneous – if a gas supply to the meter has insulated parts, or a water pipe is plastic, these would **not** be classed as extraneous.
Supplementary equipotential bonding conductor	+ Supplementary equipotential bonding is only really required as additional protection against fault if disconnection times cannot be met. + For example, the equipment is on the end of a long circuit and the bonding links that equipment to any extraneous equipment – this ensures equal potential for the longer fault duration.

Topic 3.4 Types of conductive parts

REVISED

When considering ADS and how earthing and bonding are used to protect against the risk of electric shock, it is important to understand the two categories of conductive parts that may need earthing or bonding.

The two types are shown in Table 4.8.

Table 4.8

Exposed conductive part	+ Conductive parts that form part of the electrical installation but would not normally be live.
	+ They can become live if an earth fault happens, including:
	  + steel conduit
	  + steel trunking
	  + steel tray
	  + metal accessories, such as switches and socket-outlets
	  + metallic equipment or appliances, such as luminaires.
	+ Exposed conductive parts must be connected to earth by a CPC.

Extraneous conductive parts	+ Conductive parts that are not part of the electrical installation but provide a path to earth, including:
	  + metallic service pipes (gas, oil, water)
	  + steel duct work
	  + structural steel.
	+ They must be connected to the electrical installation by bonding. This is to ensure equipotential under fault conditions.
	+ It is very important to note that if the part does **not** provide a path to earth, it is **not** extraneous but still may be metallic. If something is insulated from earth, it is not electrically a danger.
	+ The danger would be if the part has a path to earth and someone touched it while in contact with a part made live due to a fault.
	+ This is because the extraneous part provides a path for current to flow. If the part was insulated, the path would not exist and so current would not flow.
	+ If an insulated part was connected to bonding, this adds a danger that wasn't there in the first place!

Topic 3.5 Types of earth fault paths

REVISED

Disconnection times for ADS systems of protection against electric shock are completely affected by the total earth fault loop impedance of the system (Z_s).

It is the Z_s that dictates the fault current:

$$\text{Earth fault current} = \frac{U_0}{Z_s}$$

Where:
+ U_0 is the nominal voltage to earth and this is normally 230 V.
+ Z_s is the total system earth fault loop impedance.

The total earth fault loop impedance is made up from several parts in the system, which are as follows:
+ Z_e: The impedance of the supply network line and earth return path external to the installation.
+ R_1: The resistance of any line conductor up to the point of the fault.
+ R_2: The impedance of any CPC or earthing conductor forming the earth return path back to the means of earthing – includes the installation's main earthing terminal (MET).

The IET On-site Guide gives maximum values of Z_e for different earthing arrangements. These are not required by any regulations, but are values the DNOs try not to exceed.

They are:
+ 0.35 Ω TN-C-S
+ 0.8 Ω TN-S
+ 21 Ω TT.

Figure 4.12 shows the total earth fault path for the three common earthing arrangements. The difference for each is the external earth return path. The diagram shows the following external paths:
+ Path 1: TN-C-S
+ Path 2: TN-S
+ Path 3: TT

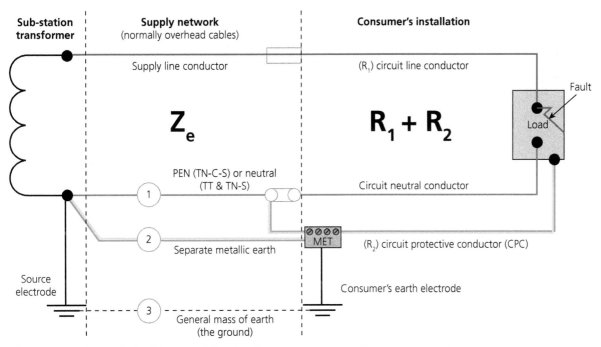

Figure 4.12 The earth fault loop path for the three common earthing arrangements

A TT installation does not have a fixed metallic earth return and relies on the resistance of the general mass of earth – this means the total Z_s is likely to be high and unreliable. Therefore, TT installations should be protected at the incoming supply to the installation by an RCD, as these disconnect at much lower fault currents (which are typical with higher impedances).

To calculate the total earth fault loop impedance, the following formula is used:

$$Z_s = Z_e + R_1 + R_2$$

The values of $R_1 + R_2$ can be determined using Tables I1 and I3 of the IET OSG.
+   **Table I1** gives values of resistance, in milli-ohms per metre (mΩ/m), for different combinations of cable sizes for line and CPC conductors at 20°C.
+   **Table I3** gives a factor needed to adjust the resistance from 20°C to the cable operating temperature. This is commonly a factor of 1.2 for thermoplastic PVC cables that operate at a maximum temperature of 70°C.

Exam tip

Remember that the total earth fault loop impedance is the sum of all the parts of the path, so they must all be added together.

Worked example

A circuit is wired using a multi-core, 2.5 mm² line and 1.5 mm² CPC copper conductors with thermoplastic PVC insulation. The circuit has a length of 22 m where it supplies a fused connection unit. The supply and installation form a TN-S earthing arrangement where the Z_e has been measured and recorded as being 0.4 Ω.

Calculate the total earth fault loop impedance for the circuit under operating conditions.

Looking at Table I1 of the OSG, the value in mΩ/m for a 2.5/1.5 combination is **19.51 mΩ/m** at 20°C.

Table I3 gives a factor (f) for thermoplastic PVC cable of **1.2** as the CPC is incorporated in the multi-core cable. So:

$$R_1 + R_2 = \frac{m\Omega/m \times L \times f}{1000} = \Omega$$

So:

$$R_1 + R_2 = \frac{19.51 \times 22 \times 1.2}{1000} = 0.52 \ \Omega$$

And:

$$Z_s = Z_e + R_1 + R_2$$

So:

$$Z_s = 0.4 + 0.52 = 0.92 \ \Omega$$

Now test yourself   TESTED ◯

6   A circuit is wired using a multi-core, 4.0 mm² line and 1.5 mm² CPC copper conductors with thermoplastic PVC insulation. The circuit has a length of 34 m where it supplies a socket-outlet. The supply and installation form a TN-C-S earthing arrangement where the Z_e has been measured and recorded as being 0.09 Ω.

Calculate the total earth fault loop impedance for the circuit under operating conditions.

Table 7.1 in the IET OSG gives guidance on the maximum lengths of circuits for different circuit types and ratings. This helps you to make sure that the maximum earth fault loop impedances are not exceeded, as this would increase the disconnection times.

LO4 Understand the principles of final circuits

In this Learning Outcome we look at the different types of circuit, as well as basic methods for looking at load capacity and voltage drop.

Topic 4.1 Arrangements of final circuits

There are two types of socket-outlet final circuit layout:
1 Radial
2 Ring

> **Final circuits** Circuits that supply socket-outlets or current-using equipment, unlike distribution circuits, which supply distribution boards.

Radial power

Radial power circuits are very simple direct circuits that supply one or more items of current using equipment such as fixed items of electrical equipment or socket-outlets.

Table H2.1 in the IET On-site Guide gives some guidance on the cross-sectional areas (CSA) for live conductors of these circuits based on the floor area served. The current-carrying capacity of the conductors is based on the protective device rating. So, from Table H2.1, an A2 radial is based on a 32 A device rating and an A3 radial on a 20 A device rating.

Radial power circuits that supply one dedicated item of equipment are often referred to as whatever they are supplying, such as:
+ a shower circuit
+ a cooker circuit
+ an immersion heater circuit
+ a space heating circuit.

These are still radial circuits in the way they are wired – but their current-carrying capacity and protective device ratings are based on the load they are supplying.

Radial lighting

Lighting circuits are also wired as radials but with the exception of having switching control parts to the circuit. The lighting circuit can be configured in the following ways:
+ **Three-plate at luminaire:**
  + Where the circuit supply is taken to the luminaire and the switch line and return are tapped off.
  + This method means the luminaire will always have a permanently live connection.
  + This is suitable where emergency luminaires are required.
+ **Conduit method:**
  + This is where single core cables are used.
  + The switch is supplied first and this controls the luminaire.
+ **Three-plate at switch:**
  + This is one of the most common ways to wire lighting circuits.
  + The supply cables are taken to the switching positions and all luminaires controlled via the switch are wired from the switch.
  + This reduces the amount of connections at a luminaire where space is tight.

Lighting can be controlled or switched in several ways:
+ **One-way:** where one switch controls one or more luminaires. This requires $1 \times$ one-way switch.
+ **Two-way:** where two switches control one or more luminaires.

+ **Two-way and intermediate**: where three or more switches control one or more luminaires. This requires $2 \times$ two-way switches and as many intermediate switches as required.

Another type of lighting circuit is an extra-low voltage (ELV) lighting circuit, or **SELV** circuit. This is where an isolating transformer is supplied at 230 V. This is reduced to an output voltage of 12 V to supply ELV lighting.

Remember that when voltage is reduced, current increases – this will have an effect on cable sizes.

> **Check your understanding**
>
> 9 What is the maximum rating of protective device, permitted by BS 7671, for a lighting circuit supplying several lighting points having Edison screw lamp holders?

> **Now test yourself** TESTED ⬤
>
> 7 There are two luminaires in a room and both are rated as being 50 W. One is a 230 V luminaire and the other is a 12 V luminaire. How much current does each take?

Ring

Ring circuits used to be very common in domestic installations to supply a number of socket-outlets. They are a good method of spreading loading around a circuit, as the conductors of a ring are wired in parallel. This helps with current capacity as well as volt drop.

However, the use of ring-final circuits is decreasing due to equipment becoming much more energy efficient with less of a need to spread loading.

One disadvantage of a ring-final circuit is that all conductors are wired in parallel. If a conductor became an open circuit, this would not be detected, as the circuit would still operate.

This situation means the remaining conductors could become overloaded as the current is no longer shared. The earth fault loop impedance could double, increasing disconnection times.

Because ring-final circuits share the current demand between two conductors, the cable current-carrying capacity is based at 20 A per conductor, rather than the 32 A protective device rating.

Appendix H and Table H2.1 in the IET OSG give further guidance on ring-final circuits.

Topic 4.2 Factors that affect load capacity

 REVISED ⬤

At Level 3, you will learn the full cable design procedure. But at this level, we keep it simple and base the selection of conductor sizes using the tables in the IET OSG.

Before selecting, we do need to determine two circuit values:
+ design current (I_b); and
+ nominal rating of protective device (I_n).

To calculate the design current, we use one of the following methods:
1 **Where a load has a power rating**:

$$I_b = \frac{\text{watts}}{\text{volts}}$$

2 **Where the circuit is for socket-outlets**:

$$I_b = I_n$$

Once design current has been determined, the nominal rating of protective device can be selected using:

$$I_n \geq I_b$$

Calculate the design current (I_b) and suitable rating of protective device for a 230 V single-phase 3 kW water heater.

$$I_b = \frac{3000}{230} = 13.04 \text{ A}$$

As I_b = 13.04 and:

$$I_n \geq I_b$$

The next suitable size of protective device is a 16 A rated device to protect the circuit.

So I_n = 16 A.

Now test yourself   TESTED

8   Calculate the design current and nominal rating of protective device for a 230 V, single-phase 4.2 kW heater.

Once device ratings have been selected, the circuit can simply be determined using Table 7.1 of the IET OSG, based on the protective device ratings and type of circuit (ring or radial).

The table will give a range of cable cross-sectional areas that can be used, including maximum circuit lengths based on the earthing arrangement.

Now test yourself   TESTED

9   Determine suitable circuit criteria for a 230 V single-phase radial power circuit to supply an 8 kW shower. The circuit is protected by an RCBO and the installation forms part of a TN-C-S earthing arrangement.

Topic 4.3 Factors and requirements of voltage drop

REVISED

Table 7.1 of the IET OSG takes voltage drop into consideration – any circuit selected using this table will be suitable for voltage drop requirements.

Voltage drop can also be calculated for a circuit if the actual voltage drop was needed.

Voltage drop is affected by:
+ cable resistance (based on its cross-sectional area)
+ length of the circuit
+ the design current (I_b) of the circuit.

So, you will need to know these!

The cable resistance, based on the conductor CSA, has been calculated into the amount of voltage lost in milli-volts, per ampere, per metre of circuit (mV/A/m) at operating temperatures. It is given in the following tables within the OSG:
+ F4ii for single-core cables
+ F5ii for multi-core cables.

To determine the voltage drop, apply:

$$\text{Voltage drop} = \frac{\text{mV/A/m} \times \text{Length} \times I_b}{1000}$$

Typical mistake

Some may think the value in mV/A/m should be multiplied by two to allow for the neutral conductor. The value of voltage drop given in Tables F4ii and F5ii takes into consideration the line and neutral resistances for every one metre length of circuit – so it doesn't need to be doubled.

Also, some questions include the design current and the nominal rating of protective device. Always use the design current to calculate voltage drop!

Worked example

Determine the voltage drop for a 230 V single-phase circuit wired in 2.5 mm² multi-core cable, having a length of 27 m and a design current of 14 A.

Using Table F5ii, a two-core 2.5 mm² cable has a voltage drop of 18 mV/A/m:

$$\frac{18 \times 27 \times 14}{1000} = 6.8 \text{ V}$$

Now test yourself   TESTED

10 Determine the voltage drop for a shower having a design current of 34 A and wired in 10 mm² multi-core cable to a length of 31 m.

BS 7671 states the current maximum voltage drop for circuits as being:
+ **power circuit**: 5% of the supply voltage (11.5 V for a 230 V supply)
+ **lighting circuit**: 3% of the supply voltage (6.9 V for a 230 V supply).

LO5 Understand technical information

In this Learning Outcome, we will revisit the information available when planning, designing, installing and commissioning an electrical installation. Some of the information we have seen already as it relates to the health and safety activities we covered in Unit 201.

Topic 5.1 Guidance publications used for electrical installation   REVISED ●

There are many publications available to help give you information on installing an electrical system. One you will be familiar with is the IET On-site Guide (OSG). Remember, you are allowed to take this into your exam with you.

Other publications available from the IET include:
+ Guidance Note 1: Selection and erection
+ Guidance Note 2: Isolation and switching
+ Guidance Note 3: Inspection and testing
+ Guidance Note 4: Protection against fire
+ Guidance Note 5: Protection against electric shock
+ Guidance Note 6: Protection against overcurrent
+ Guidance Note 7: Special locations
+ Guidance Note 8: Earthing and bonding
+ IET Guide to the building regulations.

Other guidance publications available include the 'Electrician's Guide to Good Electrical Practices', which is very similar to the OSG.

The Health and Safety Executive (HSE) also produces guidance for all aspects relating to health and safety at work. These are covered in more detail in Unit 201 on pages 12–13.

Topic 5.2 Regulations that apply to electrical systems

There are three main documents that an electrician will need to reference or work to when they are installing electrical systems:

1 The Electricity at Work Regulations (EWR)
2 BS 7671: The IET Wiring Regulations: requirements for electrical installations
3 Building Regulations

Electricity at Work Regulations

In Unit 201, we looked at these statutory regulations in depth, as they give legal rules that electricians must follow when working on electrical systems. These include regulations relating to the following areas (others are applicable):

+ **Regulation 4: Systems, work activities and protective equipment**
  + States that all electrical installations must be safe and maintained.
+ **Regulation 5: Strength and capability of electrical equipment**
  + No electrical equipment shall be put into use where its strength and capability may be exceeded in a way that might be dangerous.
  + This is why inspection and testing are undertaken.
+ **Regulation 10: Connections**
  + Where necessary to prevent danger, every joint and connection in a system shall be mechanically and electrically suitable for use.
+ **Regulation 12: Means for cutting off the supply and for isolation**
  + Relates to safe isolation and providing a means of locking off systems and circuits.
+ **Regulation 13: Precautions for work on equipment made dead**
  + States that work should never be undertaken on live parts, unless there is absolutely no reasonable alternative.
+ **Regulation 14: Work on or near live conductors**
  + Gives requirements where working live or near live parts must take place.
+ **Regulation 15: Working space, access and lighting**
  + States that there must be adequate space and light to carry out any electrical work.

Building Regulations

In the UK, the Building Regulations vary depending on the nation you are working in. The Building Regulations (England) are very similar to the Building Regulations (Wales), but Scotland and Northern Ireland have Building Standards, which are different.

The Building Regulations or Building Standards are statutory documents, so they are legal requirements in their respective nation.

The Building Regulations (England) have the following parts that are applicable to electrical work activities:

+ **A: Construction**
  + Details areas such as depths of chases, cutting holes in joists and other structurally related tasks.
+ **B: Fire safety**
  + Details areas such as maintaining fire seals when installing systems through fire barriers or installing recessed luminaires.
+ **E: Resistance to sound**
  + Relates to the integrity of barriers (like Part B) but also the passage of sound.
+ **L: Conservation of fuel**
  + Relates to the need for energy-efficient lighting and methods of lighting control (e.g. outside lights are not left on during the day).

> **Exam tip**
>
> If a question in your exam relates to the Building Regulations and applicable parts, it will be based on the England and Wales versions, not Scotland or Northern Ireland.

109

+ **M: Access and use**
    + Relates to the location of socket-outlets and switches so they are accessible for all persons.
+ **P: Electrical**
    + Gives legal requirements for electrical systems in dwellings only, as the Electricity at Work Regulations cover all other types of building.

BS 7671

This is a non-statutory document that outlines the requirements for all electrical installations. The On-site Guide takes the relevant regulations applicable to single-phase systems up to 100 A and puts them into an easier to understand format.

Exam tip

The IET On-site Guide lists all the relevant Building Regulations in Part 1, which you are allowed to take into your exam. The list in the OSG contains further sections not shown here.

The regulations are currently divided into seven parts:
+ **Part 1 Scope and fundamental principles**
    + Details what installations the document covers.
    + Outlines the fundamental principles for safety, which are high level requirements of what needs to be considered during the design and installation stages.
+ **Part 2 Definitions**
    + Lists all the technical words and terms used in the document with definitions.
+ **Part 3 Assessment of general characteristics**
    + Details what must be assessed before any design or work begins to ensure the right design and installation are provided.
+ **Part 4 Protection for safety**
    + Details very technical requirements that need to be put into place to ensure installations are designed and installed to protect against risks.
    + Risks include electric shock, fire, overloads and electromagnetic interference.
+ **Part 5 Selection and erection**
    + Details the requirements for choosing equipment and how it should be installed.
+ **Part 6 Inspection and testing**
    + Details the requirements for the initial verification process for new installations or circuits.
    + Details the requirements for periodically testing existing installations.
+ **Part 7 Special locations**
    + Outlines further requirements for installation or locations where there are specific risks due to the environment.

In addition to the parts of BS 7671, there are Appendices that provide detailed information to assist in meeting the regulations.

Topic 5.3 Manufacturers' information to support planning of electrical activities

REVISED

Manufacturers provide information on their products. These include the following items:
+ **Data sheets**
    + Specific information relating to the product (e.g. dimensions, suitable environmental conditions and electrical data such as voltage, current ratings etc.).
+ **Installation instructions**
    + Includes instructions on how to fix and set up the equipment.
    + Outlines circuit conditions, such as the need for RCD protection on their products.

- **Product warranties**
  - Should always be passed on to the client in case something goes wrong with the equipment during the warranty period (which is usually 12 months).

It is worth noting that manufacturer's instructions override any requirements in BS 7671. For example, circuit conditions may not require RCD protection under the requirements of BS 7671 – but, if the manufacturer's documentation states that RCD protection is required, the RCD **must** be installed.

Topic 5.4 Drawings used to plan electrical activities

REVISED

In electrical installations, there are many different types of drawing you may come across for different purposes. These are covered in Table 4.9.

Table 4.9

Plans/layout drawings	<ul><li>Used to show the layout of a building or site.</li><li>Include the intended location of the equipment.</li><li>They would always be drawn to scale (see Topic 5.5).</li></ul>
Schematic plans	<ul><li>Show a sequence of control, e.g. distribution systems in a large building indicating what isolators control which distribution boards.</li><li>Include details such as cable sizes and ratings/type of isolator.</li></ul>
Circuit diagrams	<ul><li>Show detailed connections, specific terminals and how they are connected to each other.</li><li>Examples include systems such as boiler control circuits and which connection on the thermostat is connected to a particular motorised valve terminal.</li></ul>
Wiring diagrams	<ul><li>Show what cable connects to the components.</li><li>For example, a one-way switch controlling one luminaire would have a two-core cable connecting them.</li><li>This would be shown with a single line but a description of the cable type (e.g. 6242Y for two-core twin and CPC cable).</li></ul>
Block diagrams	<ul><li>Similar to schematic plans – they include a sequence but tend to show how components or items of equipment interact with each other, rather than a sequence of control.</li></ul>

Typical mistake

Wiring diagrams and circuit diagrams often become confused as many manufacturers produce hybrid versions of the two. However, if individual conductors are shown throughout the drawing, including which terminals they connect to, it is a circuit diagram.

Topic 5.5 Symbols and scales used in electrical documents

REVISED

There are no standards that need a particular electrical drawing symbol to be used. The drawing should be clear if it has a key or legend that describes each symbol. Some symbols used are contained in BS 8541, but these mainly cover symbols used on circuit diagrams.

Drawing scales are only applicable to layout/plan drawings, as the other drawings are not intended to show exact positions – only connections.

Drawing symbols

Wiring diagrams and block diagrams do not typically use symbols but instead use labelled shapes to indicate the component. Figure 4.13 shows a sample of the different types of symbols used in circuit, schematic or plan drawings.

	Plan symbol	Circuit or schematic symbol
One-way double-pole light switch	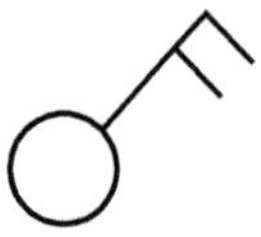	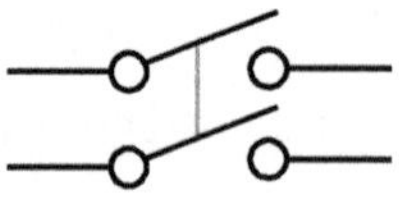
2-gang socket-outlet with single-pole switch	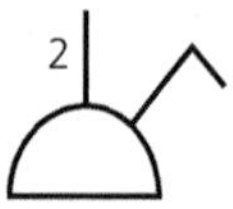	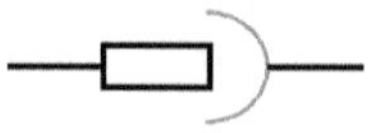
Switch disconnector	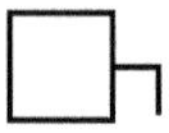	

Figure 4.13 Sample of drawing symbols

Symbols for lighting switches are always circular as this is the standard switch symbol. The single stalk shows it is one-way, but the two lines at right angles to the stalk show the switch is double-pole.

Looking at the samples in Figure 4.13, the number next to the symbol for a socket-outlet means the socket is 2-gang or that it has two outlets. The stalk, which is at an angle, shows the socket-outlet is switched. The single line at right angles on the stalk shows the switch is single-pole.

The switch disconnector plan symbol is a standard symbol for isolators and switch-fuses used for distribution systems. But the circuit diagram symbol has a circle on it, which means the switch is designed for on-load switching.

Drawing scales

Because drawings need to represent large areas, buildings or sites, but on small pieces of paper, scales need to be used.

Scales are chosen for two reasons:
1 To show a suitable level of detail.
2 To suit the paper size for the drawing.

Commonly used drawing scales are shown in Table 4.10.

Table 4.10

Scale	What the scale represents on the drawing	What the scale represents on site (in real life)	Examples of applications
1:1	1 mm	1 mm	A one-to-one drawing is an exact size and no conversion is required.
1:10	1 mm	10 mm	Every 1 mm on a drawing represents 10 mm on site, so a wall on a drawing measuring 178 mm would be, on site: $$178 \times 10 = 1780 \text{ mm or } 1.78 \text{ m}$$
1:50	1 mm	50 mm	Every 1 mm on a drawing represents 50 mm on site, so a wall on a drawing measuring 178 mm would be, on site: $$178 \times 50 = 8900 \text{ mm or } 8.9 \text{ m}$$

Check your understanding and progress at **www.hoddereducation.co.uk/myrevisionnotes**

Scale	What the scale represents on the drawing	What the scale represents on site (in real life)	Examples of applications
1:100	1 mm	100 mm	Every 1 mm on a drawing represents 100 mm on site, so a wall on a drawing measuring 178 mm would be, on site: $$178 \times 100 = 17,800 \text{ mm or } 17.8 \text{ m}$$
1:500	1 mm	500 mm	Every 1 mm on a drawing represents 500 mm on site, so a wall on a drawing measuring 178 mm would be, on site: $$178 \times 500 = 89,000 \text{ mm or } 89 \text{ m}$$

Check your understanding

10 What would be the length of a line on a 1:50 drawing which is to represent a wall which is 17 m?

11 What are the relationships between the following commonly used paper sizes?
  + A4
  + A3
  + A1

LO6 Understand requirements for obtaining and providing client information

This Learning Outcome focuses on the information which is passed to the client, such as financial or technical information.

Topic 6.1 Types of financial information

REVISED ○

If a business is to do well, it is important to understand the financial documentation used. Table 4.11 lists some basic documents and why they are used.

Table 4.11

Document	Use
Estimates	+ Give the client a good idea of the cost of work – but not a fixed price as there may be unforeseen issues. + For example: an estimate is given to re-wire a house where existing cables appear to be in conduits in a wall so there shouldn't be a need to chase the wall. But when work begins, it is clear that the top part of the conduit cannot be easily accessed so the wall does need chasing, leading to extra costs. If a fixed price was given, this cost would have to be absorbed by the electrician or company.
Quotations	+ Fixed price costs to carry out work. + It is strongly advised that a full survey of the work is done to ensure there are no hidden problems that may arise, before giving a quotation.
Payment schedules	+ Provide the client with details on when to provide payment for work on a regular basis, rather than at the end of a project. + This is because some contracts can take a long time to complete. + Payments may coincide with the completion of phases of the work or it may be on a timed basis (e.g. monthly).

Document	Use
Invoice statements	✦ Provide the client with details of what payments have been invoiced and what payments have been made by the client and received by the contractor. ✦ Often coincide with an agreed payment schedule, so work is invoiced over several intervals.
Cancellation rights	✦ Once a quotation or estimate has been agreed, there is a statutory period of 14 days where the client (or contractor) can pull out of the agreement. ✦ They can decide not to go ahead with the work, assuming the work hasn't started. ✦ Any upfront payments of deposits would need to be reimbursed to the client.

> **Invoice** A documented request for money in return for services or goods; normally has payment terms such as 30 days to pay the money owed.

(NB: As you progress through Level 3, this information becomes more specific to larger contractual documentation.)

Topic 6.2 Types of handover information

REVISED ●

A handover is where all work is complete; testing and commissioning have been done; and the contractor who did the work is formally handing over all documents and instructions for the work completed.

Handover would also involve training or demonstrating to the client how the system operates. This may include showing how timers are set or explaining what maintenance is required at particular intervals.

The information that may be handed to the client is listed in Table 4.12.

Table 4.12

Design philosophy statement	✦ A summary of what the intention of the installation is. ✦ For example, it might state that the intention of the installation is to provide a safe means of delivering power in an energy-efficient form. ✦ This statement might be why the client chose the organisation to do the work. It is often revisited at the handover stage to show how the philosophy has been met.
As fitted drawings	✦ Apply to more complex installations where plans were issued for the work to be done. ✦ By issuing them, it is a record of what equipment or accessories have been installed, their locations and other general information (e.g. what circuit they are on).
Product user instructions	✦ It is important that all manufacturers' information for the use of any equipment is handed to the client for future referencing.
Maintenance schedules	✦ Apply to larger installations where regular maintenance tasks are required at regular intervals. ✦ The schedules will contain the tasks needed and how often they should be performed. ✦ For example, RCDs require testing for function by pressing the test button; this is required every six months.
Building Regulations approvals	✦ Some work requires approval under the Building Regulations (unless the work is done by a member of a competent person scheme). ✦ The Local Authority building controls department will issue the client with certification of conformity once the work is done – assuming the work has been done correctly!
Electrical test information	✦ A test and inspection certification is handed to the client once electrical work is completed. ✦ It proves that the installation has been handed to the client in a safe and suitable state. ✦ Clients must hold onto this information as the test data in it will be required for comparison when future periodic or maintenance testing is carried out. ✦ If the data isn't available, this could end up costing the client financially (i.e. a full survey needs to be done).
Energy performance certification	✦ Many local authorities set energy efficiency targets for new building projects, such as housing having efficient lighting or heating. ✦ Part L of the Building Regulations requires lighting to be energy efficient. ✦ The certification that comes with the appliances or equipment will need to be kept as proof that the requirements have been met.

Exam checklist

+ Methods of generating electricity
+ How electricity is transmitted and distributed to the consumer, including equipment used
+ Arrangement at the supply intake position for an installation
+ Constructional features of a consumer unit
+ Types of earthing arrangements
+ Types and functions of protective devices
+ Purpose and types of isolation and switching
+ Purpose and provision of basic protection
+ Purpose and provision of fault protection

+ Purpose of earthing and bonding
+ Types of conductive part requiring earthing and bonding
+ Parts of the earth fault loop path
+ Types and arrangements of final circuits
+ Factors affecting load capacity of a circuit
+ Factors affecting volt drop
+ Knowing the statutory, non-statutory and guidance documents used for electrical work
+ Knowing drawings, scales and symbols
+ Documentation and information issued to a client

Exam-style questions

1 Which is a renewable source for generating electricity?
   a) Nuclear
   b) Coal
   c) Oil
   d) Wind

2 Which of the following is a transmission voltage?
   a) 400 TV
   b) 400 MV
   c) 400 kV
   d) 400 V

3 What is the maximum permitted residual current rating for an RCD providing additional protection?
   a) 30 mA
   b) 100 mA
   c) 300 mA
   d) 500 mA

4 Where does the earthing conductor for a TT earthing arrangement connect the MET to?
   a) The supply PEN conductor
   b) The supply cable sheath
   c) The consumer's earth electrode
   d) The sub-station source electrode

5 What type of fault current would not be detected by a circuit breaker?
   a) High impedance residual fault current
   b) Low impedance short circuit current
   c) Low impedance earth fault current
   d) High current overloads

6 What does the term 'breaking capacity' mean?
   a) The maximum current a fuse can safely disconnect
   b) The maximum current the fuse can normally carry
   c) The minimum current needed for quick disconnection
   d) The minimum current causing overload operation

7 Which protective device will operate at low residual current, overloads and high fault currents?
   a) MCCB
   b) MCB
   c) RCD
   d) RCBO

8 A circuit is correctly co-ordinated so a local protective device disconnects before the main service fuse when a fault occurs in an appliance.
   What is this co-ordination called?
   a) Diversity
   b) Selectivity
   c) Determination
   d) Suitability

9 Which is a method of fault protection?
   a) Barriers
   b) Enclosures
   c) ADS
   d) Basic insulation

10 What does IP2X represent when considering basic protection?
   a) Stopping fingers touching live parts
   b) Stopping 1 mm wires touching live parts
   c) Causing disconnection in 2 seconds
   d) Causing disconnection at 2 amperes

11 What is the maximum permitted disconnection time for a 32 A distribution circuit forming part of a TN-S earthing arrangement?
   a) 0.2 seconds
   b) 0.4 seconds
   c) 1 second
   d) 5 seconds

12 What conductor is connected to extraneous conductive parts?
   a) Circuit protective conductor
   b) Earthing conductor
   c) Main protective bonding conductor
   d) Circuit line conductor

13 Which of the following is an extraneous conductive part?
   a) Metal case of an oil tank
   b) Metallic dimmer switch
   c) Metal wet system radiator
   d) Metallic structural supports

14 What earthing arrangement is shown in Figure 1?

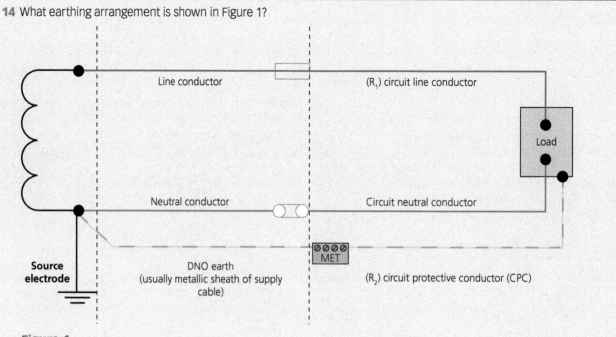

Figure 1

a) TN-C-S   c) TT

b) PME   d) TN-S

15 Which of the following formulae is correct for calculating Z_s?

a) $Z_s = Z_e - (R_1 + R_2)$ c) $Z_s = Z_e - (R_1 - R_2)$

b) $Z_s = Z_e + (R_1 + R_2)$ d) $Z_s = Z_e + (R_1 - R_2)$

16 What is the $R_1 + R_2$ for a circuit, at operating temperatures, where both the line and CPC conductors are wired in multi-core 4 mm² copper conductors to a length of 15 m?

a) 0.13 Ω   c) 0.22 Ω

b) 0.17 Ω   d) 0.27 Ω

17 What does the symbol in Figure 2 represent?

Figure 2

a) one-way SP switch

b) two-way SP switch

c) one-way DP switch

d) two-way and intermediate switch

18 What is the voltage drop for a single-phase circuit having a design current of 17 A, a length of 27 m and wired in multi-core 6 mm² copper cable?

a) 3.35 V   c) 6.90 V

b) 5.35 V   d) 11.5 V

19 Which of the following documents is statutory?

a) BS 7671   c) IET GN3

b) EWR   d) IET OSG

20 A drawing has a scale of 1:50 and a wall measuring 26 m is to be represented by a line on the drawing.

What is the length of the line in mm on the drawing?

a) 130 mm

b) 260 mm

c) 520 mm

d) 840 mm

Check your understanding and progress at **www.hoddereducation.co.uk/myrevisionnotes**

Exam breakdown

Test specifications

This qualification is assessed using a range of tests, including a synoptic assignment and a written examination. Your exam has a test specification. The test specification shows the following information:
+ assessment method (e.g. multiple-choice or written)
+ examination duration
+ permitted materials (e.g. closed or open book)
+ number of questions.

The way the **knowledge is covered by each test** is laid out in the exam specification below:

Exam specification

The assessment objectives (AOs) are the types of question that could be asked across the exam and the exam is spread across the AOs, as detailed below.

Assessment objective	020/520 weighting (approx. %)
AO1 Recalls knowledge from across the breadth of the qualification.	42%
AO2 Demonstrates understanding of concepts, theories and processes from across the breadth of the qualification.	38%
AO4 Applies knowledge, understanding and skills from across the breadth of the qualification in an integrated and holistic way to achieve specified purpose.	20%

The way the exam covers the content of the qualification is laid out in the table below.

Assessment type: Multiple-choice exam

Duration: 120 minutes

Assessment conditions: Invigilated examination conditions

Grading: Pass/Merit/Distinction

020/520	Duration: 2 hours		
Unit	Title	Marks available	Weighting
201	Health and safety and industry practices	4	7
202	Electrical science	18	30
203	Electrical installation	12	20
204	Electrical technology	14	23
n/a	Applied knowledge and understanding	12	20
	Total	**60**	**100**

The exam draws from across the mandatory content of the qualification, using the following types of questions:
+ **Multiple-choice questions:** These questions confirm your breadth of knowledge and understanding (AO1/AO2).
+ **Multiple-choice applied knowledge and understanding questions:** These questions give you the opportunity to demonstrate higher level, integrated understanding through application, analysis and evaluation (AO4).

Applied knowledge questions will integrate several areas from the test specification into one question. For example, rather than a question asking a simple AO1 recall question asking for the correct ladder ratio, a **stretch** question would also ask for the length of the ladder needed to climb a wall a given height. This will require recall of the correct ladder ratio, as well as understanding of the rules of trigonometry, and applying them to this given situation.

Permitted exam materials

You are permitted to take into your exam the following materials:
+ IET On-site Guide
+ BS 7671.

While you are allowed to take in BS 7671, you probably will not need it, as at Level 2 any research questions rely on tables in the On-site Guide.

You are allowed to place tabs on your pages to allow you to quickly find certain information or tables. It is recommended you do this, but do not overdo it, as the tabs can become unreadable or confusing!

Using the practice exam-style questions in this book will give you a really good idea of which tables are often referenced. As a sample, these include the following:
+ **Table 7.1**: shows conventional circuit selection information
+ **Table D3/4**: gives the maximum support spacings for conduits/trunking
+ **Tables E1–E6**: give cable and spacing factors for trunking and conduit.

Examiner's tips: Preparing for the exam

+ Make sure you revise before the exam. Use practice exam questions, such as the ones in this book, to identify areas of weakness and focus your revision on these.
+ Check the test specification to see what areas the exam focuses on. Check the duration and number of questions in the exam.
+ Be sure you know what time the exam starts. The exam begins at that time, not the time you arrive!
+ Arrive at least 30 minutes before the start time so you can settle, check you have everything you need and ensure your calculator works.
+ Make sure you know the requirements for the exam. Is the exam open book? What books can you take in? Do you need a calculator? Always take in pens or pencils and ask for scrap paper from the invigilator when you go in. It always helps to write things down if you need to perform calculations or jot down key information from a structured question.

Examiner's tips: During the exam

+ Listen carefully to the instructions given to you by your invigilator before you start the exam.
+ Attempt to answer every question. Leaving a multiple-choice question unanswered will be marked as incorrect. Have an educated guess as you have a 25% chance of being right.
+ Read the question and, for multiple-choice, every answer before you attempt the question. If you read it too quickly, you will miss key words or information.
+ Do not rush. You have, on average, two minutes per question.
+ Never let a question get to you! If a single question is not working out or you cannot find an answer, '**flag it**', have a guess and move on. If one question unsettles you, the remaining questions will become much harder because your state of mind has changed. Save the remaining questions by moving on **before** you are unsettled. If you flag the question, you can easily go back if you have time at the end. Sometimes, further questions may prompt you towards the answer of the question you were struggling with.

Check your understanding and progress at **www.hoddereducation.co.uk/myrevisionnotes**

Question structures

Your exam is a multiple-choice question paper which is taken using a computer. You will be given a question stem then four options. Only one of the options is the correct answer. The other three options may be very close to the answer, but something will make them incorrect.

Read every question at least once, including all of the optional answers, before attempting to answer the question.

Closed questions

Most of your exam will consist of closed questions. These include a clear and precise question where the question ends with a question mark.

Open questions

Very occasionally, an open question is used where the correct answer completes the sentence started by the question. These are rare but they may be used when quoting a specific technical term or guidance from the permitted publications.

For example (taken from section 3.6.2 in the IET On-site Guide):

RCDs for additional requirements for socket-outlets can be omitted in:

a) non-domestic premises where a documented risk assessment determined it unnecessary

b) a domestic installation where a documented risk assessment is carried out first

c) any installation where the socket-outlets are fitted at high level locations

d) industrial installations where the socket is located outdoors.

The correct answer is a).

Negative questions

Although these are not often used, beware of negative questions. These are questions that ask you to find an option which is **not** suitable or applicable. For example:

Which item of equipment is **not** required to secure a circuit breaker following checking for isolation?

a) Padlock

b) Unique key

c) Proving unit

d) Locking clamp

The correct answer is c) because all the other items are used to lock the circuit breaker in the off position (secure) but a proving unit is used to check the voltage indicator works correctly.

Structured questions

Some questions will be based on a common scenario. The same scenario will reappear in each question and the questions will build from an AO1-style recall question to a complex AO4 applied knowledge question. Make sure you answer these questions in relation to the specific scenario rather than to general situations.

Command words

Command words that start a question are not as important in a multiple-choice question as they are in a written exam, where they give a good indication of depth in a written exam. However, multiple-choice questions might have command words that emphasise what you need to answer – these words will be in **bold**. These words include terms such as:

+ **maximum**: e.g. what is the highest, biggest, or longest ...
+ **minimum**: e.g. what is the lowest, smallest, shortest ...
+ **most**: normally followed by 'suitable', 'appropriate', 'common', and is normally linked to a procedure
+ **not**: used in negative questions.

The Evolve platform

All City & Guilds multiple-choice online tests are taken using the Evolve platform. Information about the Evolve platform can be found on the City & Guilds website.

The synoptic assessment

The synoptic assessment is a way of testing your understanding, skills and behaviours over several practical and design tasks. These change each year and are based around a scenario such as an office or shop. You are not marked against each task; instead you are marked on the way you perform over all the tasks.

The synoptic assessment usually involves the following five tasks:

1 Basic design and planning based on the given scenario.
2 Safe isolation procedure.
3 Erection and use of access equipment.
4 Install wiring systems and components to form a small installation.
5 Self-reflection where you reflect on each task and describe what went well and what you could improve, given the chance again.

Below is some advice for preparing for Task 1.

Task 1 Design and planning

You will need to apply skills and understanding in Task 1 of your synoptic assessment. These may include the following:

+ Using **scaled drawings** to apply scales and measurements to work out quantities of the necessary materials. You will be provided with an A3 drawing to scale from accurately. Be sure you remember how to apply scales, such as 1:50 or 1:100 etc.
+ Create a **detailed material list** applicable to the installation shown in the scaled drawing. The materials sheet also requires you to justify your choice of materials. You need to demonstrate your understanding here by using technical terms and understanding of installation conditions. For example:

Quantity	Materials	Justification/Reasons for choice
12	2-gang metal-clad 13 A socket-outlets	Metal-clad has been chosen as the installation is a workshop and these sockets are good where impact is likely.
3	20 A 30 mA RCBOs Type B	As the RCBOs are protecting general equipment, Type B gives best protection and 30 mA RCD provides the required additional protection.
2	2-way 1-gang metal-clad switches	These are suitable for the workshop environment and allow switching from two places. They only need to be 1-gang as only one switch is needed.

+ Complete a **circuit schedule** where you will need to divide the installation into circuits. Consider the following points:
    + How many lights should be on one circuit? If a fault occurs, will all lighting be lost?
    + What equipment is likely to be used and plugged into the socket-outlets? How many should be on one circuit?
    + What are the lengths of the circuits and will they be affected by voltage drop? What table in the On-site Guide will help?
    + How will the circuits be adequately protected? What rating of protective device is needed and what size cables will co-ordinate with this?
+ Produce a **method statement** for the safe isolation procedure. Remember, a method statement is a document that describes how to do a task in the correct order. Make sure that you know the procedure, including the following:
    + Is any specific equipment needed?
    + What specifically needs testing?
    + Is permission needed before switching off?
    + What exactly is being switched off?
    + Does anyone need to be made aware, as isolation may create dangers such as lifts stopping or vital equipment being turned off?

Always remember: what you produce in Task 1 may be moderated by a City & Guilds moderator. The following documents from Task 1 are submitted as evidence and may be checked:
+ Your circuit schedule
+ Your materials list with justifications
+ Your method statement for safe isolation
+ Your drawings, showing any workings out for the scales and measurements.

Remember, your synoptic assignment isn't just marked on what you produce but also the way you produce it. What behaviours did you display? Were you methodical, did you give attention to detail, were you efficient with research, or did you constantly ask for help and reassurance?

Being prepared sufficiently for your synoptic assessment will help you to display the right behaviours.

Glossary

Basic protection A technical term used in BS 7671 – refers to the insulation around live parts, or the barriers and enclosures housing live parts, which prevent users from touching live parts. Page 17, 93

Bespoke Something that has been made for a particular client or use to a particular specification. Page 80

Bi-metallic strip Two different metals bound together. Each metal expands at a different rate, which causes the strip to gradually bend with heat. The bending movement is used to activate a switch. Page 94

BS EN 60309 The standard for socket-outlets for varying current voltages and current ratings. They may be 16 A, 32 A, 45 A or higher. Their colour denotes their voltage rating, such as yellow for 110 V, blue for 130 V and red for 400 V three-phase. Page 79

Builder's board A small but safe temporary supply. It usually consists of a wooden board near the incoming electrical supply. Mounted on the board is a distribution board fed from the incoming supply, with a range of socket-outlets to provide a power source while the building's systems are out of operation. Page 27

Cable containment systems Systems that completely surround the cable(s), such as trunking. Page 68

Cable management systems Systems that support the cable(s), such as tray. Page 68

Charge (symbol Q) The measure of electron flow in a material; it is measured in coulombs (C). Page 42

Chase A channel cut into a masonry wall to a particular depth to lay in cables or boxes so they will be buried in the wall and not seen. Page 67

CHIPS Stands for Chemical Hazard Information and Packaging for Supply Regulations. Page 24

Conductors Materials, such as gold, silver and copper, which have atoms less densely packed together and so allow electron flow. Page 42

Continued professional development (CPD) Electricians are expected to keep up to date with the latest technology advancements, as well as changes to regulations. Page 33

Current (I) The measure of charge over time (t) and measured in amperes (A). So:

$$I = \frac{Q}{t} \ (A).$$ Page 43

Current-using equipment A term used in BS 7671 for any appliance, load, luminaire etc. that uses current in order to function, or converts electricity into another form of energy (e.g. light or heat). Page 79

Design current The full load current of the circuit under normal operating conditions. Page 90

Dissipated To have used energy or power. Page 50

Down-times The duration a system is not operating due to fault or failure. Sometimes loss of supply can be very expensive for businesses. Page 75

Draw tape Draw tape is normally nylon and rigid enough to be pushed through sections of conduit, where cables are then tied to it and pulled through by pulling the draw tape back again. Page 76

Drawn in Cables are pulled through conduits. This requires a lot of pre-planning to make sure the right number of cables are pulled into each section. Page 76

Dressing Means making sure cables are neat and straight without kinks or twists. Page 76

Earth Also known as 'terra-firma' from the French word for 'earth' – 'terre'. Many international electrical regulations are written in French, so 'T' is the symbol for 'earth' when looking at earthing arrangements. Page 17, 91

Eddy current A current that is induced into a metal by the magnetic field. These currents rotate around the metal (like eddy currents in a stream causing mini whirlpools) and these currents cause the material to heat up. As using energy to produce heat is a loss, reducing eddy currents reduces this energy loss. Page 58

Electro-magnetism Where a magnetic force is produced by passing current through a conductor. When a direct current is passed through a coiled conductor, this produces the same effect as a bar magnet with magnetic poles. Page 51

Equal potential Where two parts have the same voltage, which actually reduces the risk of shock. If two surfaces were live to 230 V and someone touched them both, the voltage is the same across the person, so the potential difference is 0 V and harmful current will not flow. Page 101

Exposed conductive parts These are metal parts that have electrical parts but would not normally be live; but they could become live if a fault happens (e.g. metal casings of appliances or metal trunking and conduit). Page 93

Extraneous conductive parts Metal parts that have a potential path to earth but are not part of the electrical system, such as metal gas pipes. Page 101

Ferrous metal A metal containing iron, which because of this can stick to a magnet. Metals that do not contain iron, such as aluminium, are known as 'non-ferrous' metals. Page 45

Final circuits Circuits that supply socket-outlets or current-using equipment, unlike distribution circuits, which supply distribution boards. Page 105

Final temperature The point where the insulation starts to melt. Page 45

Flush accessories Items, such as switches or socket-outlets, where all that is seen on the surface is the face plate. The rest is buried in the wall. Page 67

Former The shaped part of the bender that the conduit rests in when it is bent to keep its shape. Page 63

Fossil fuels Resources such as coal, oil or gas that are mined from the earth and burnt to produce heat – this produces carbon greenhouse gas pollution. They are not renewable resources. Page 84

Hazard Something that is dangerous and could cause harm (e.g. working at height). Page 8

Health and Safety Executive (HSE) The UK body responsible for shaping and reviewing health- and safety-related regulations, producing research and statistics, and enforcing the law. Page 8

INDG455 An industry guidance document published by HSE on the safe use of ladders and stepladders. INDG guides are free to view on the HSE website. Page 13

Indices A method of simplifying a large or small quantity by using a power of 10; for example:

$$6 \times 10^{12} = 6,000,000,000,000.$$ Page 38

Induced To force the creation of (in electrical terms), i.e. a current will force the creation of a magnetic field, and so a magnetic field will force the creation of a current. Page 51, 57

Inspection boxes Conduit boxes where access can be made to cables and cables can be drawn in at these points. Page 70

Insulators Materials, such as plastics or glass, that have atoms which are densely packed together, and so electrons are less likely to flow from one atom to another. Page 42

Invoice A documented request for money in return for services or goods; normally has payment terms such as 30 days to pay the money owed. Page 114

IP protection An international numerical code system for the levels of protection against the ingress of dust or foreign bodies (represented by the first number) or water and moisture (represented by the second number). Page 100

IP2X The barrier or enclosure has no hole bigger than 12.5 mm in diameter, so fingers cannot access the inside of a barrier or enclosure. The 'X' means that protection against water or moisture is not applicable in this situation. Page 100

IP4X A specific electrical code which means there is no hole with a diameter greater than 1 mm. Page 100

IPXXB A specific electrical code which means that if a finger enters the enclosure, any live parts are spaced more than 80 mm from the hole, so a finger length could not touch the live part. Page 100

Laminated Where a block or object is made up of many layers. In the case of a transformer, these layers are bolted together to form the core. Page 58

Masonry Material such as brick cement, concrete or plaster. Page 67

Maximum operating temperature The temperature above which the insulation resistance begins to break down, allowing current to flow with the circuit voltage. Page 45

Means of earthing The means of earthing for an installation is the earth provided by either the DNO, for a TN arrangement, or the consumer's earth electrode for a TT arrangement. Page 101

Mechanical aids Items used to make handling materials easier and safer, such as trolleys, sack barrows and dollies. Page 22

Mechanical damage Damage such as tears, cuts, abrasion, crushing (or similar). Page 17

Method A set way to do something; a method statement is the method in writing. Page 18

MICC A mineral-insulated copper-clad cable. The mineral used for the cable insulation performs well at very high temperatures, making this cable suitable for high temperature applications (e.g. fire alarm systems where performance needs to continue in the event of a fire). Page 64

Mutual induction Where one winding produces a magnetic field and that magnetic field cuts through a second winding, producing a current in that second winding. Page 57

Non-statutory Not law but following them is considered as best practice. Page 8

Ohm's Law The relationship of current, voltage and resistance in an electrical circuit: $V = I \times R$. Page 43

Peak demand When more electricity is being used around the whole country, such as cold, dark days when more heat and light are needed. Sometimes, this can be at half time during a major football game on TV, when everyone boils a kettle to make a cup of tea! Page 85

Platforms Defined by the Working at Height Regulations as any surface above ground height used as a place of work or as means of access to a place of work and includes scaffold, suspended scaffold, cradles and mobile platforms. Page 15

Pliable Means flexible or bendable. In this case it means flexible. Page 74

Potential difference The difference in voltage from one terminal to another. Page 51

Premature collapse When the wiring system falls down due to fire before parts of the building structure supporting it. Page 78

Renewable resource A resource that can be used repeatedly and does not run out because it is naturally replaced, such as wind or solar (sunlight) energy to generate electricity. Page 84

Resistance The measure of how well a cable or insulator conducts electricity in ohms (Ω). The lower the value of ohms, the better it conducts; the higher the ohms, the worse it conducts. Page 43

Risk How likely a hazard is to cause harm and how much harm it could cause. Page 8

Segregation A technical term used in BS 7671 meaning to keep different systems apart from each other to reduce the risk of interference or danger where different voltages are used. Page 69

Self-induction Where one winding produces a magnetic field and that magnetic field induces a current back into the same winding. Page 57

SI units The international system for units of measurement; these are the base values that should be used in any formula. Page 36

Solenoid A conductor wound into a tight helix to produce an intensified magnetic field. Page 53

Stand-off A sturdy attachment to a ladder that enables the ladder to rest standing off from a wall around 0.5 m, leaving the top of the ladder clear from resting against brittle surfaces such as gutters or windows. Page 15

Statutory The regulations are law and must be followed. Page 8

Sub-station The final transformer before the consumer – they are found in many different places, such as behind fences, at the end of roads or inside brick buildings. Page 87

Toolbox talks Regular, on-site, informal safety meetings. Some sites have them daily before work starts. Page 21

Toroidal Circular or doughnut-shaped. Page 94

Transpose To rearrange a formula in order to make the value you need to find the subject of the formula. Page 39

Turbine A turbine is basically a series of propellers which are turned at very high speeds by high-pressure steam or water. Page 85

User check A term used in the IET Code of Practice for in-service inspection and testing of electrical equipment (CoP ISITEE), which is considered a vital safety precaution before using any electrical equipment. Page 18

VDE A European testing institute that gives insulated tools their safety certification to rigorous standards of safety. Certified VDE tools carry the VDE mark (see Figure 3.2). Page 65

Weir A low dam across a river that increases the force of the water as it flows over the top. Sections of a weir can be raised or lowered to regulate the force of the water. Page 85

White goods Appliances used in the home, usually the kitchen, such as washing machines, fridges and dishwashers. Page 33

Young person Someone aged between 16 and 18, but may include 15 if they turn 16 in that academic year. A young person under the age of 16 is considered a child. Page 19

Check your understanding and progress at **www.hoddereducation.co.uk/myrevisionnotes**